ANAL SURGERY FOR GENERAL SURGEONS

A HANDBOOK OF BENIGN COMMON ANO-RECTAL DISORDERS.

NORMAN A BLUMBERG

AuthorHouse™
1663 Liberty Drive
Bloomington, IN 47403
www.authorhouse.com
Phone: 1 (800) 839-8640

Published by AuthorHouse 05/28/2015

ISBN: 978-1-5049-0619-7 (sc)
ISBN: 978-1-5049-0657-9 (e)

Library of Congress Control Number: 2015906100

Print information available on the last page.

Norman A Blumberg MB BCh MD FRCS FACS

Clinical Associate Professor Department Surgery U T Houston Texas

Formerly Chief of Surgery Bayshore Medical Center Pasadena Texas

Formerly Specialist Surgeon Johannesburg South Africa

Formerly Specialist Surgeon to the Mines Benefit Society Johannesburg

Formerly Consultant Surgeon Coronation Hospital and University of Witwatersrand Johannesburg.

Formerly Tutor in Anatomy University of Witwatersrand Johannesburg S Africa

FOREWORD

Norman Blumberg MD BCh FRCS FACS has invited me to write a foreword to his handbook on the management of common anorectal disorders and I am delighted to do so. I have been slow in putting my thoughts on paper since I have had little exposure to the diagnosis and treatment of anal and perianal diseases apart from that received as an intern in general surgery at the Cornell University Medical College/New York Hospital. My experience is as that described by the author; the management of hind gut diseases is under the supervision of a junior resident at the end of a long day of assisting other residents in the performance of more complex surgical procedures.

Dr. Blumberg received training in general and colorectal surgery in Johannesburg, South Africa and the United Kingdom at hospitals and institutions known for their excellence. His experience as a teacher whilst serving as a consultant at Coronation Hospital and the University of Witwatersrand hospitals stimulated his interest in academic surgery which he maintained parri passu with private practice in general surgery at Bayshore Medical Center in Pasadena, Texas after immigrating to the United States more than thirty five years ago. This led to his appointment as a part-time tutor in the Department of Surgery at UT in Houston, during my tenure, when he lectured to undergraduates on diseases of the hind gut, including those featured in his work.

This book emphasizes the importance of knowledge of the local anatomy in both understanding and treating common, troublesome diseases. It also emphasizes that treatment should be simple and that diagnosis does not require a multitude of exotic tests and investigations. And most importantly, it emphasizes the need to endoscopically investigate the entire colon for the cause of rectal bleeding and not be lulled by hemorrhoids which may prove to be a decoy.

The reader is well advised to not completely excise the anal ring for advanced hemorrhoids and to exercise gentleness in stretching the anus (always horizontal, without bleeding and not to exceed four fingers). In treating chronic anal fissure division of the internal sphincter should be limited and localized. Fistulectomy should not be electively performed in the presence of an abscess. Advice is also given regarding various positions for examination as well as surgical procedures.

Norman has saved the best for last in his final chapter, where he deals with the issue of bowel hygiene and appropriate dietary concerns. He also appropriately recommends against the habitual use of narcotic containing medications such as cough and pain medicines.

This book fills a long neglected need and deserves a place on every operating surgeon's bookshelf.

FRANK G MOODY MD FACS
Professor of Surgery and Emeritus Chairman and Head of Surgery, University of Texas, Houston

INTRODUCTION

Minor diseases of the anus often produce disproportionate discomfort or pain; there can hardly be a more miserable patient than one who is plagued by chronic anal problems and few as grateful as one who is permanently cured.

As a General Surgeon with a practice largely comprising Colorectal surgery, and also as an academic teacher in Surgery and Anatomy over a span of 51 years, it was long apparent to me that the Ano-rectum is the Cinderella of General Surgery and the butt (no pun intended) of endless jokes. Teaching is often haphazard and unsystematic, with the Anal Canal added as a tail end in General Surgery curricula (again no pun intended) and either relegated to junior, inexperienced trainees, or placed at the end of an operating schedule, where it is invariably preceded by more exotic and dramatic surgical procedures. Whilst there are undoubtedly many outstanding colorectal surgeons outside of specialised institutions, and although I am unaware of any studies, I believe that most Ano-rectal surgery is performed by General Surgeons; alas, not surprisingly, often badly. At least fifty percent of my own practice encompassed colorectal surgery. Here I sometimes encountered the bad outcomes of surgery performed elsewhere, e.g. anal scarring and disfigurement, mucous leak, anal stenosis, ectropion and varying degrees of incontinence. All these made simple local hygiene impossible, causing social embarrassment or mechanical difficulty with defecation.

This work does not completely cover the field of proctology. It describes the common, mostly minor diseases and procedures which are encountered every day and it does not include more highly specialized subjects such as construction of pouches, or neo- sphincteroplasty. Such topics I regard as the domain of the highly specialized rectal surgeon with a continuous, high volume turnover of complicated surgery, not the "occasional" operator. There are many outstanding textbooks on ano -rectal surgery. This work is not intended as a textbook but is a documentation of the procedures that I employed during a fifty years lifetime in surgery, in several thousands of patients. It is a reflexion of personal experience and is intended to remove the mystique that somehow remains and to help ensure that minor anal coditions are treated and cured without preventable complications.

Notwithstanding modern technological advances, C D's, DVD's etc. as teaching tools, I believe there remains a place for the written word in pages which are readily accessible in just a few moments, to refresh, such as immediately before entering the operating room. "

And so, I have long felt an urge to write this book. After many false starts, this is the final product. I hope that it will serve as a useful practical manual for all surgeons interested in this field. Virtually every disease featured will be diagnosable by a lucid history and simple clinical examination. For the most part a multitude of exotic, expensive and sophisticated tests is not required.

My intention is not to compete with the numerous existing excellent tomes on Colorectal surgery but rather to supplement them. The ultimate purpose of this book is to hopefully prevent common minor errors of judgment and technique .

Although intended primarily for General Surgeons, I hope that it may also be found useful by undergraduate and postgraduate students, surgeons- in- training , and perhaps even by specialist Colorectal Surgeons and proctologists. My objective has, at all times, been to present information in a simple, readable and practical manner. The style is intentionally didactic and conversational. In my minds eye I visualized live tutorials and demonstration of operative techniques. By retaining original diagrams and sketches I have attempted to keep the presentation personal.

In the pages that follow I describe the simple procedures and techniques that I employed during more than half a century of surgical practice and involving several thousands of patients. Some, are modifications of well established practice, some, I believe, are original. It is my hope that they will find their way into some colleagues repertoires. "

My special thanks to Hymie Gaylis M Ch MD FRCS FRCSE* formerly of Johannesburg South Africa and later living in retirement in San Diego and Professor Frank G Moody of Houston for reviewing the manuscript and invaluable advice and suggestions and Barbara Glauber Lindenberg of Houston Texas, for endless patience and help with the computer and scanner.

* means the party concerned is now deceased

TABLE OF CONTENTS

CHAPTER 1

TOOLS OF THE TRADE

Hereunder commonly used equipment and instruments. The list is not complete and you may wish to add your own preferences.

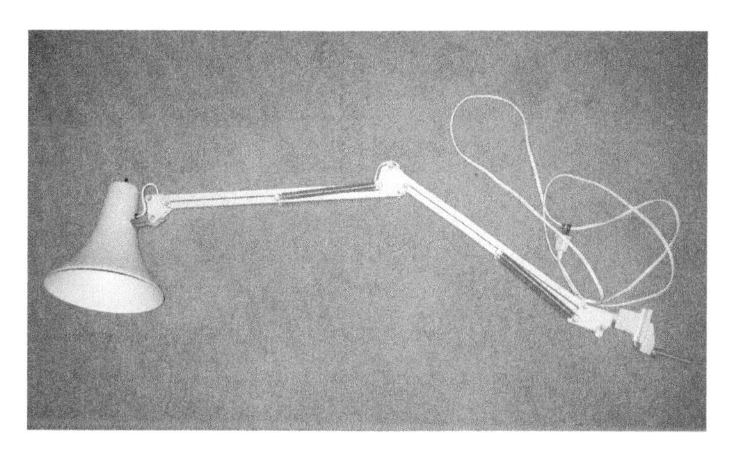

Anglepoise lamp is inexpensive, flexible and provides excellent lighting.

Magnifying operating loupes. Non flipup design.

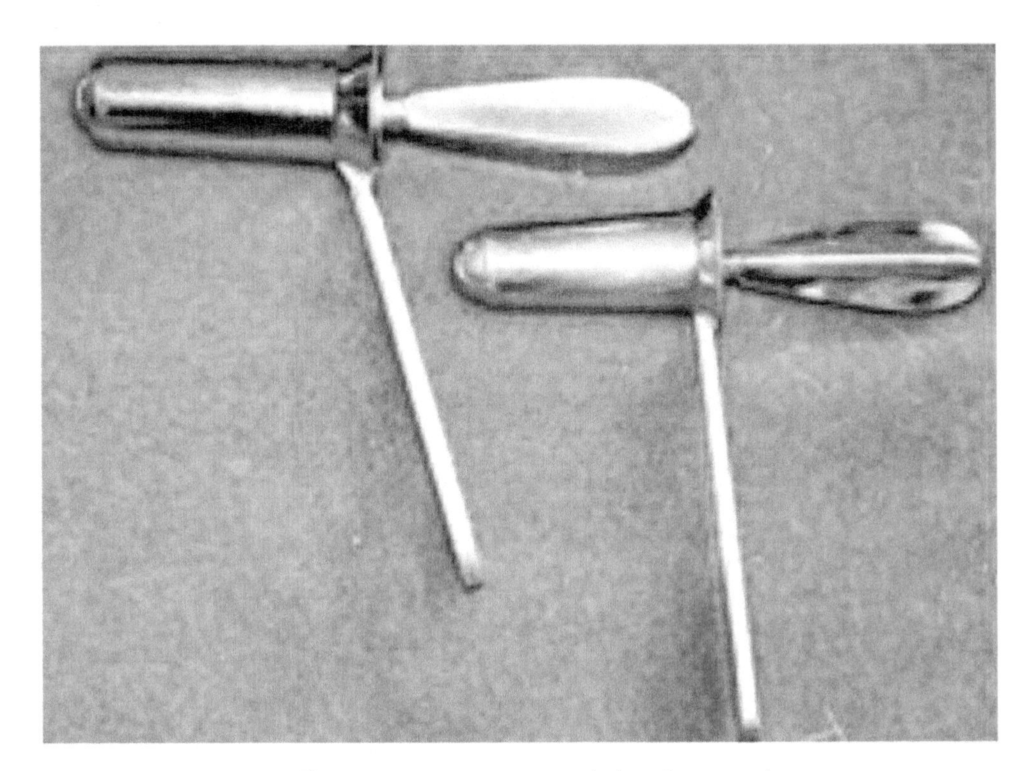

Officers pattern protoscopes (a k a Anoscopes)

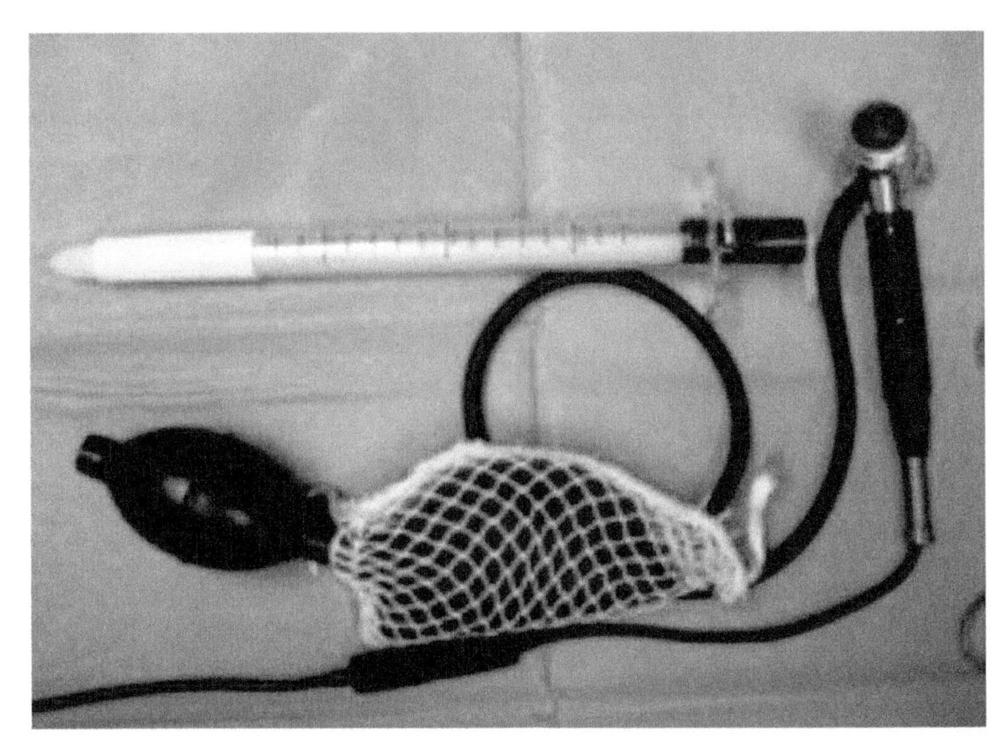

Rigid sigmoidoscope. Disposable scopes have largely replaced the traditional non disposable instruments

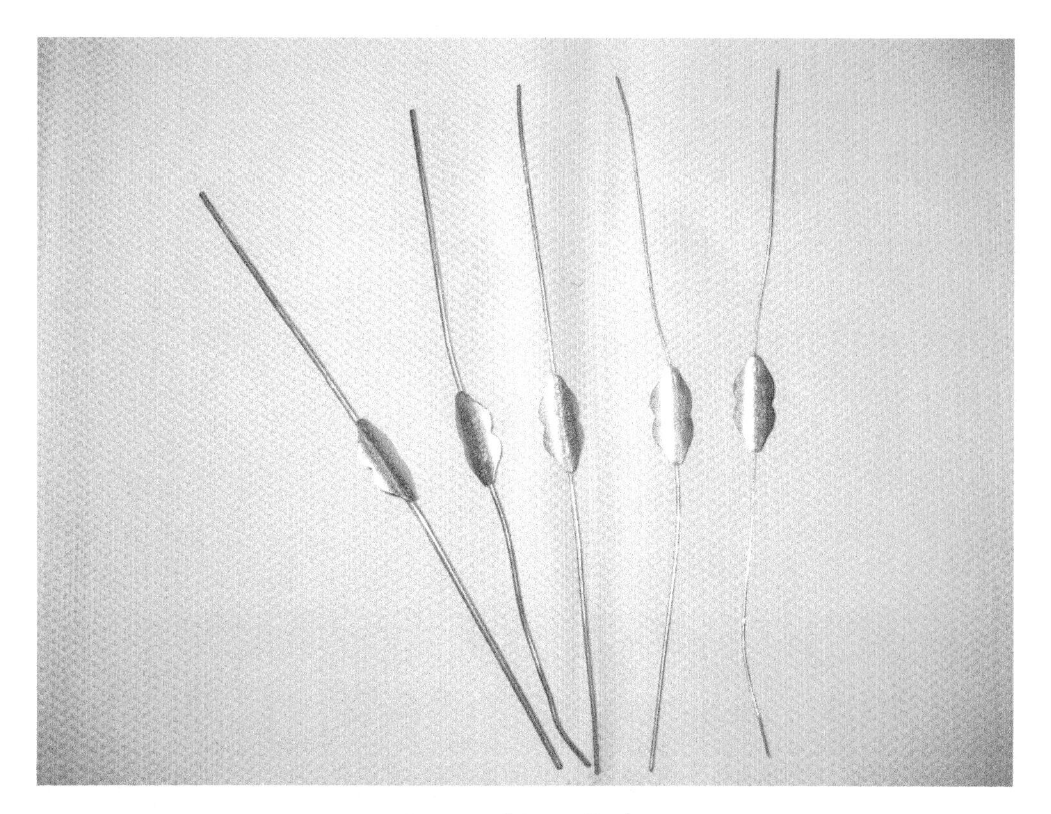

Lacrimal Duct Probes

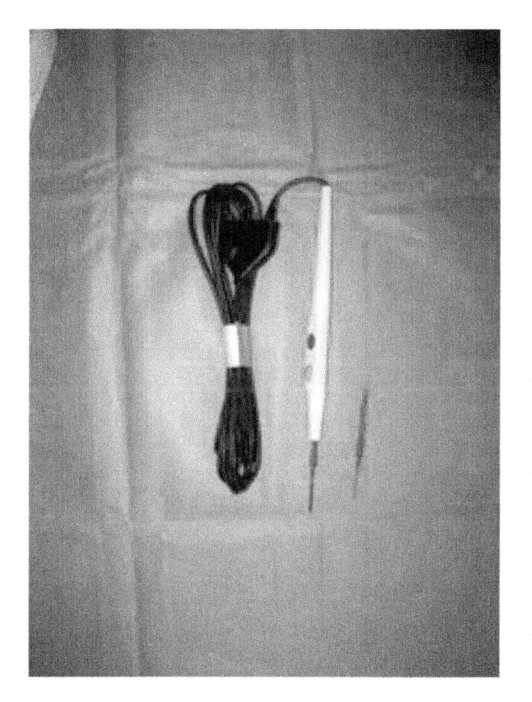

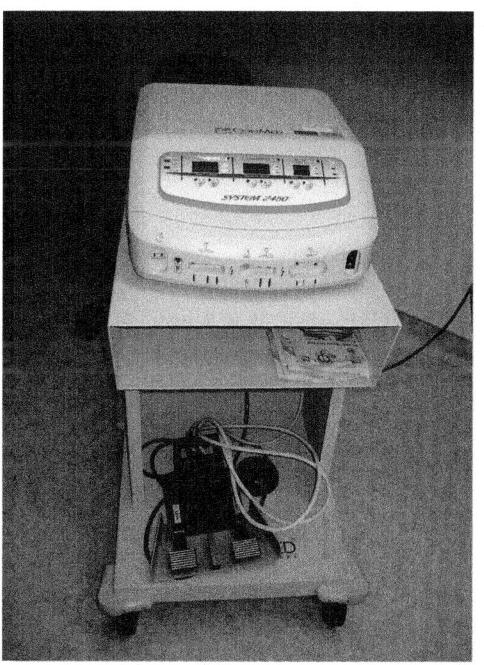

Bovie Diathermy coagulation machine

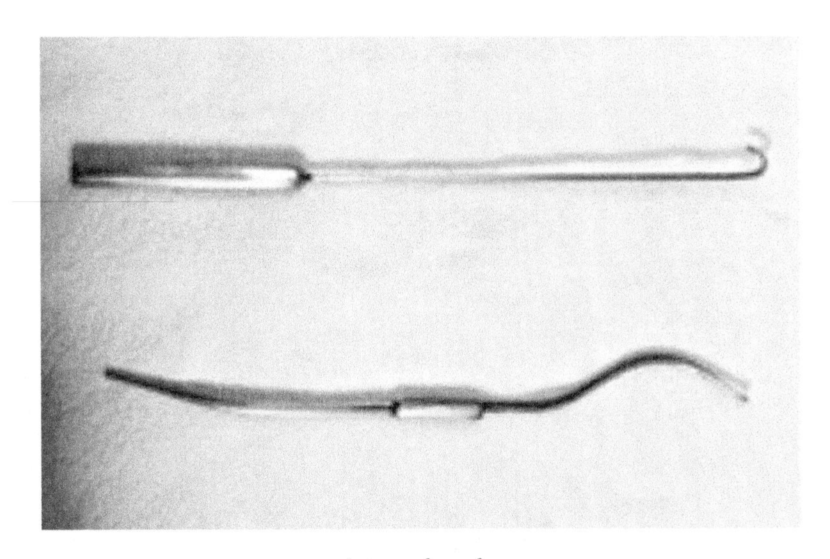

Assorted probes

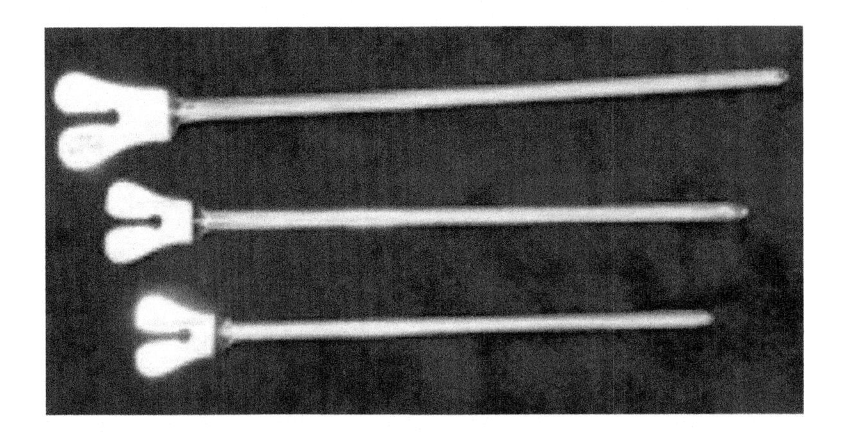

grooved directors

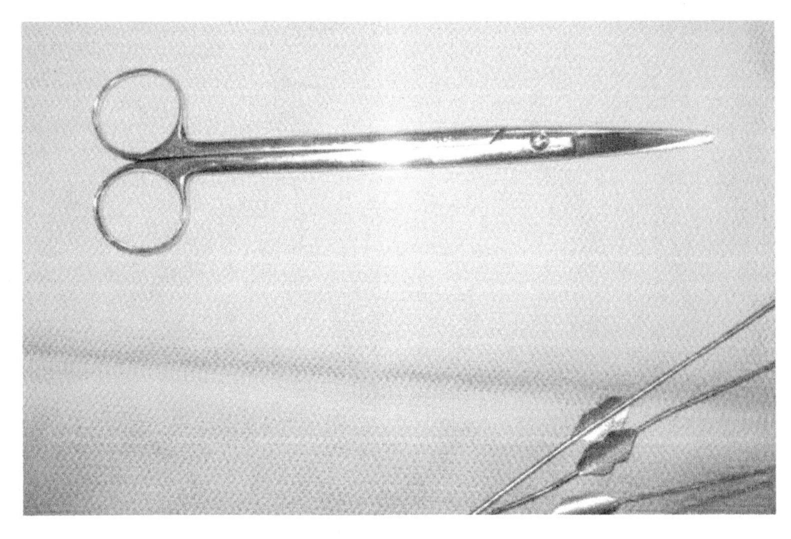

Mayo Scissors and Lacrimal duct probes

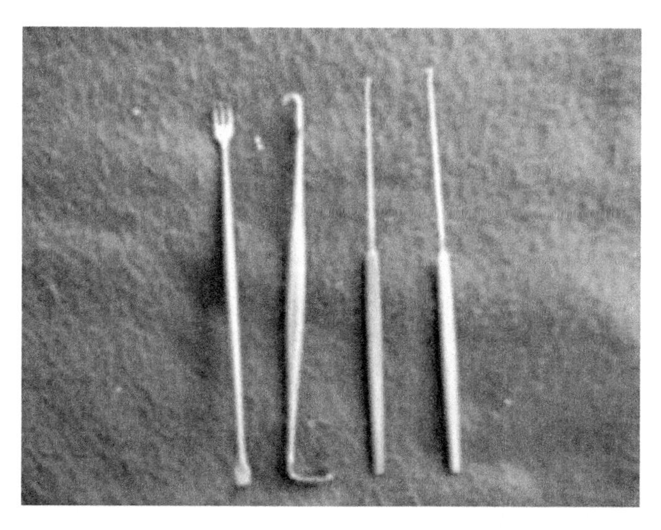

St Mary (catspaw) retractors and McIndoe skin hooks

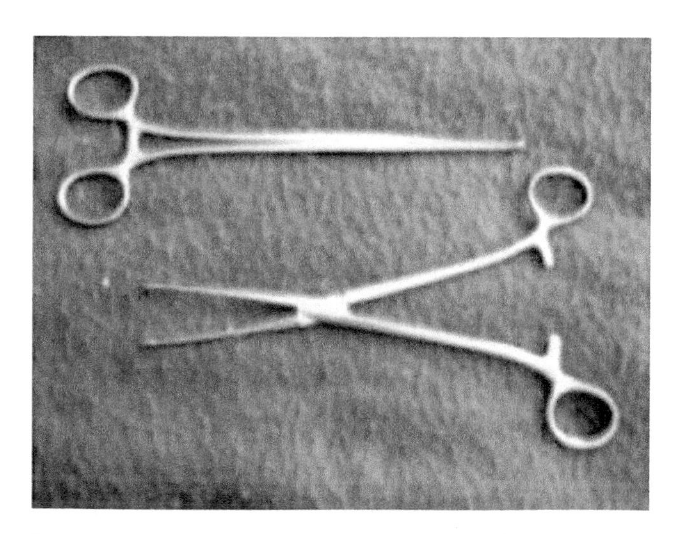

Judd Allis clamps. These have atraumatic notched ends which are non crushing

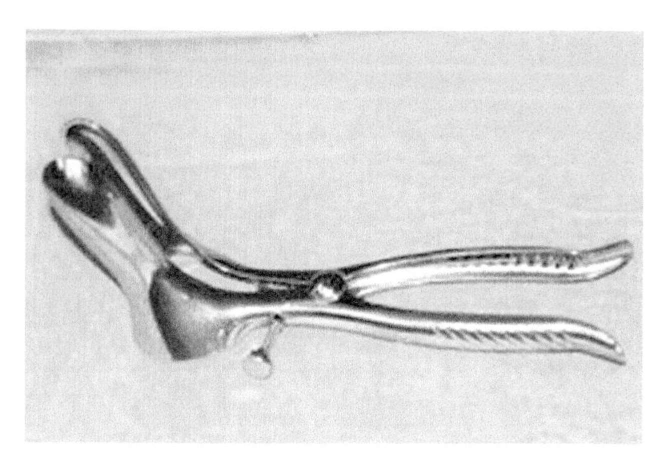

Vaired two and three bladed adjustable retractors and speculums

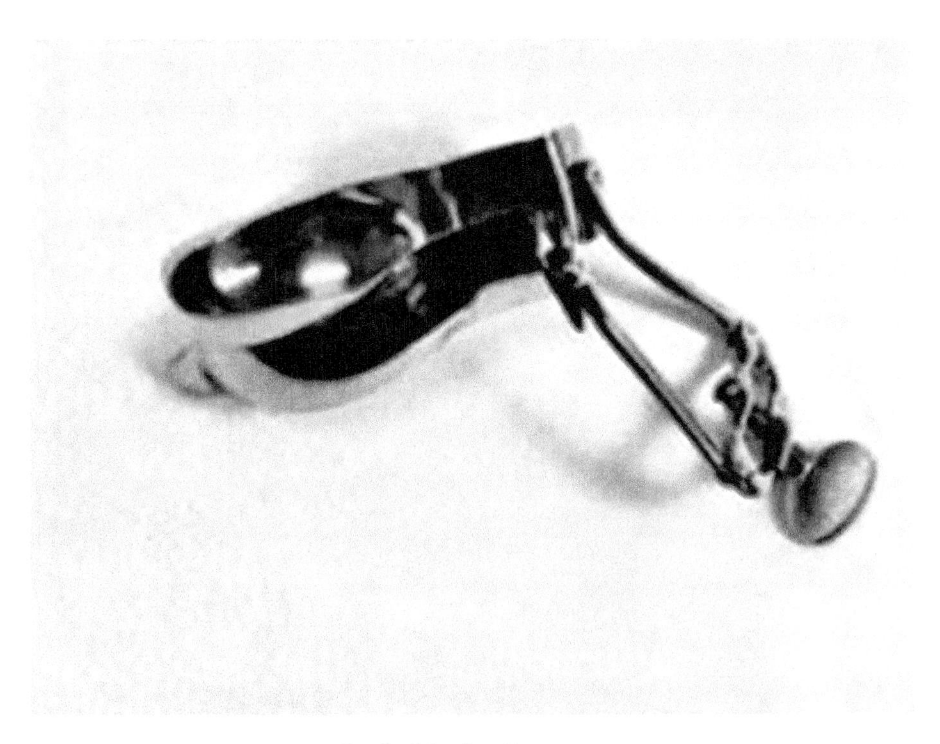

Parks Mark 1 Retractor

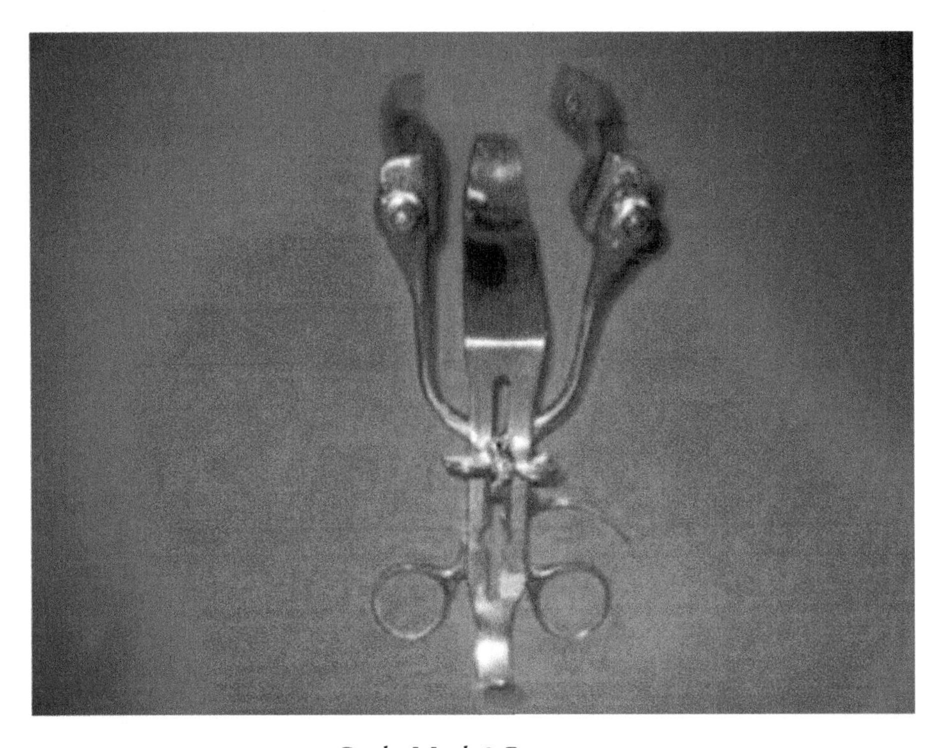

Parks Mark 2 Retractor

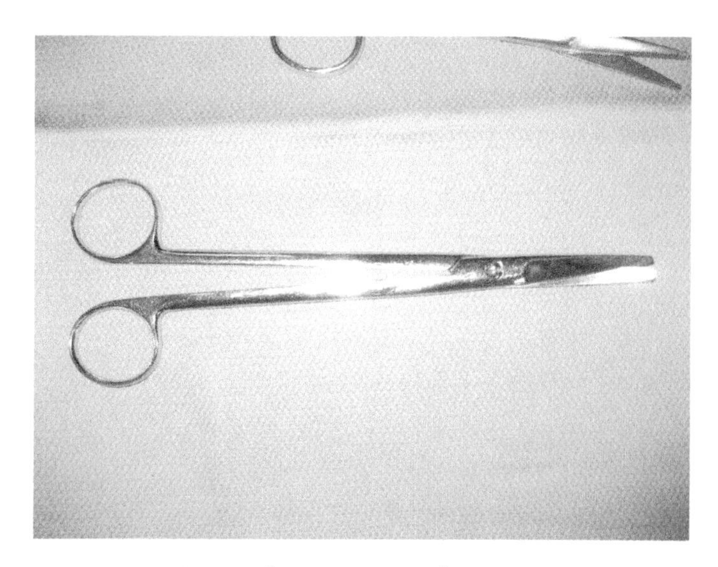

Metzenbaum or McIndoe scissors

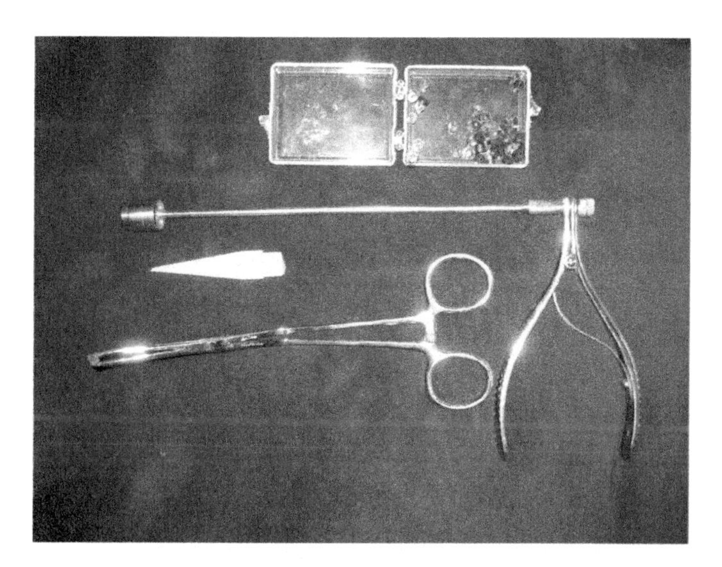

McGivney Ligator with loading cone, rubber ring plus grasping forceps

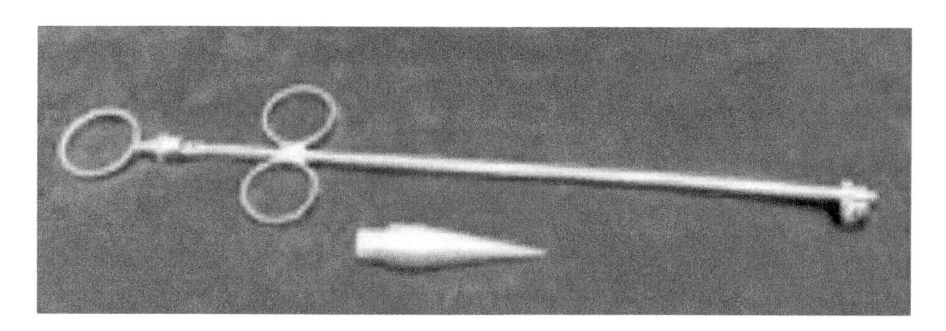

Simple Ligator (Seaward)

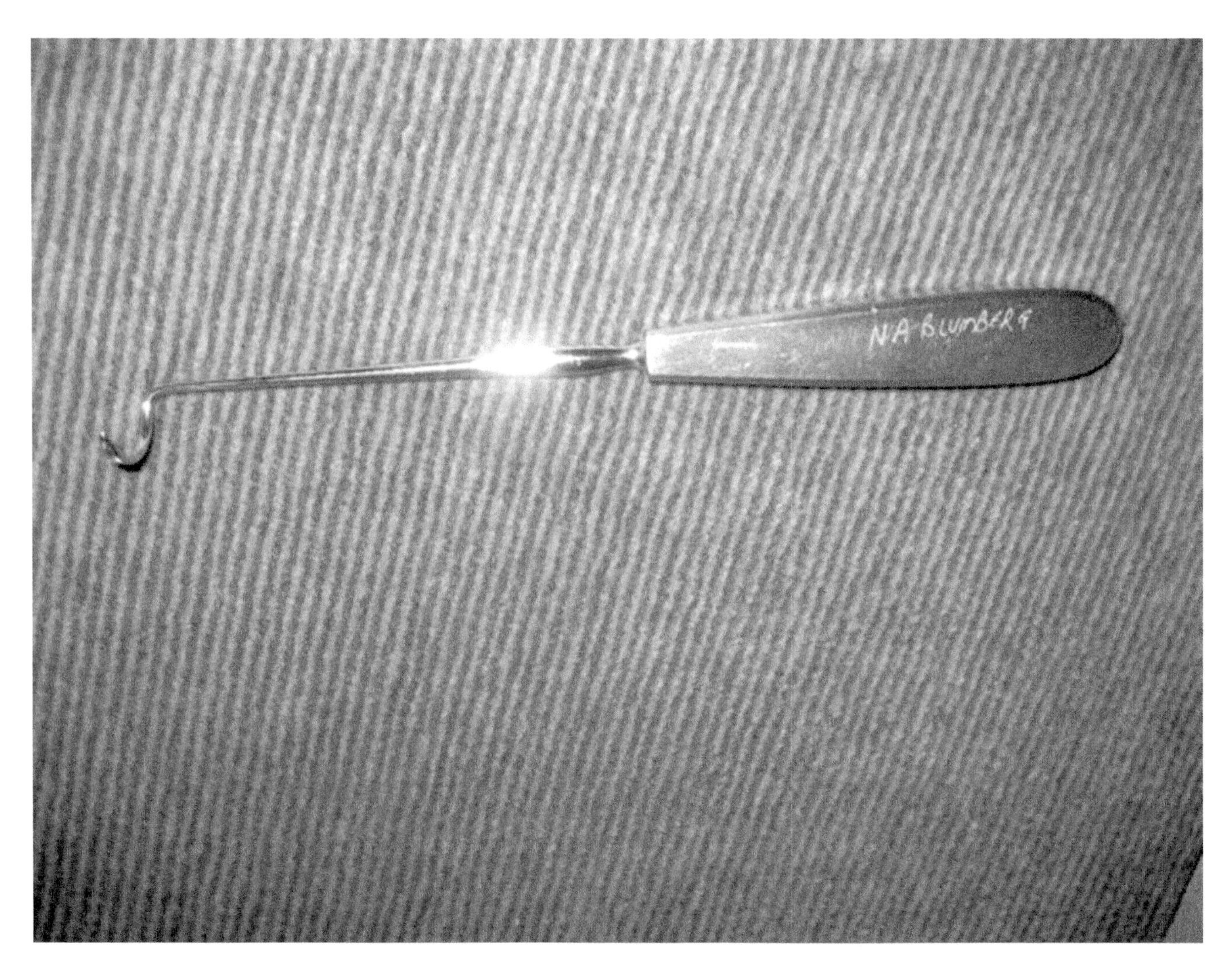

Aneurym needle, right to left

CHAPTER 2

ANATOMY OF ANO RECTUM

The Anorectum is a deceptively simple, but in reality, extraordinarily complex structure. The gross anatomy of the area is straightforward and easy to understand and remember, belying the complexity of functions relating to continence and bowel control. Although in health the system is not accorded even a perfunctory thought, when it malfunctions there can be no more miserable patient than one with even minor anorectal problems. In diagnosing and treating anal disorders there is probably no other area in the entire human body where precise knowledge of the anatomy is so critically important.

The Anorectum can be likened to a tube (the bowel and its wall) within a funnel (the surrounding muscle and sphincters).

The Anorectal Ring is a composite of smooth and striated muscle; the Internal sphincter (smooth) and the deep part of the External sphincter (striated) together with the Puborectalis component of the Levator Ani muscle (striated). The latter muscle also constitutes the pelvic floor. This complex arrangement is exquisitely fine- tuned to recognise and maintain continence for gas, liquid and solids and the ability to evacuate at will. It is innervated by both the somatic and autonomic systems and disruption of its integrity may manifest in minor or major incontinence.

The anal canal is approximately 1¼ inches long in males, slightly less in females. It commences at the Anorectal Ring and ends at the anal verge. At rest it is a narrow slit in the A/P direction. Fat masses in lateral spaces allow for distention to accommodate the bulk of stool and bowel content during defecation. The Puborectalis muscle (the most medial, innermost component of the Levator Ani) exerts a forward pull and thus angulates the Ano - rectal junction so that the anal canal runs down and backward at almost a right angle from the rectum (fig 2.1). It is postulated that this arrangement may contribute to continence.

Fig. 2.1 The pubo rectalis muscle (red) angulates the anorectal junction forward to approximately a right angle

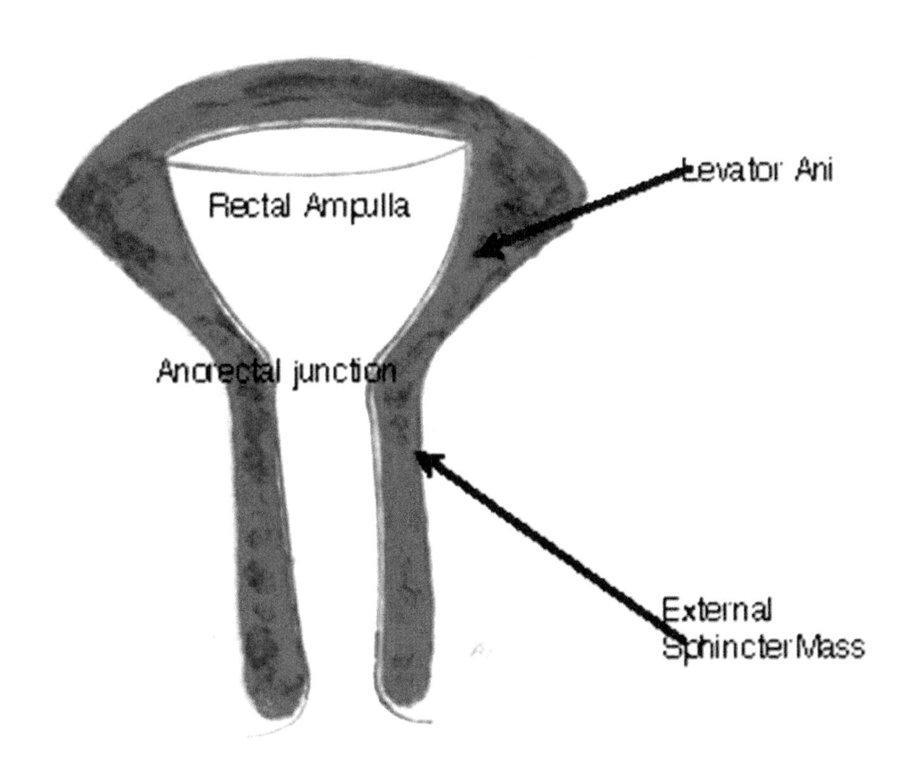

Fig 2.2

The Sphincters

The Internal Sphincter is the thickened distal end of the circular muscle of the large intestine and measures the full length of the anal canal. It is freestanding at its distal end, where it is traversed by terminal fibres of the Conjoined Longitudinal muscle. It is composed of smooth (unstriated) muscle, whitish in color. The Conjoined Longitudinal muscle fans out into widespread terminal filaments which are directed medially (and in so doing they traverse the internal sphincter), and laterally as well as inferiorly. Thereby it gains attachment to local structures, including anal mucosa and anoderm, the external sphincter and perianal skin.

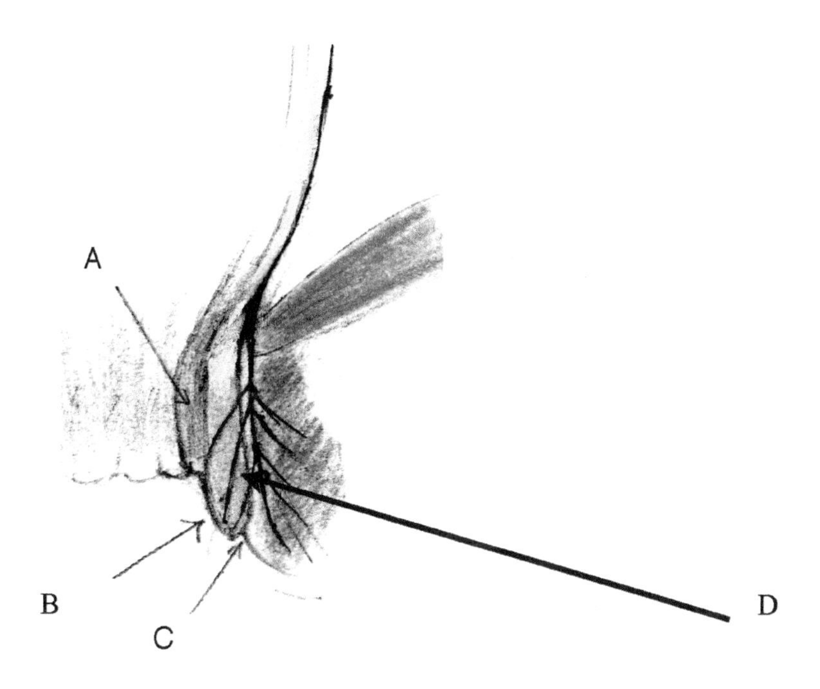

Fig 2.3 A= submucosa B= Internal Sphincter C=Palpable depression. D=The terminal fibres of the Conjoined Longitudinal Muscle (CLM) The colours are deliberately exaggerated for clarity.

The External Sphincter, red/brown in color, and composed of striated muscle, is divided into three parts. These are commonly named Subcutaneous, Superficial and Deep components. The author finds the distinction between "Superficial" and "Subcutaneous" moot and recommends that the terms Superficial (or Subcutaneous), Intermediate and Deep are less confusing and will so describe the muscles in this text.

The deep component is fused with other muscles; the Puborectalis part of the Levator Ani, longitudinal muscle and the internal sphincter. The composite forms the Anorectal ring (marking the ano- rectal junction).

Complete division of the ring will result in complete incontinence. Partial division, including of the Internal Sphincter, may result in varying degrees of incontinence. The Anorectal Ring therefore at all times deserves the greatest respect from the surgeon.

The Dentate Line (DL)

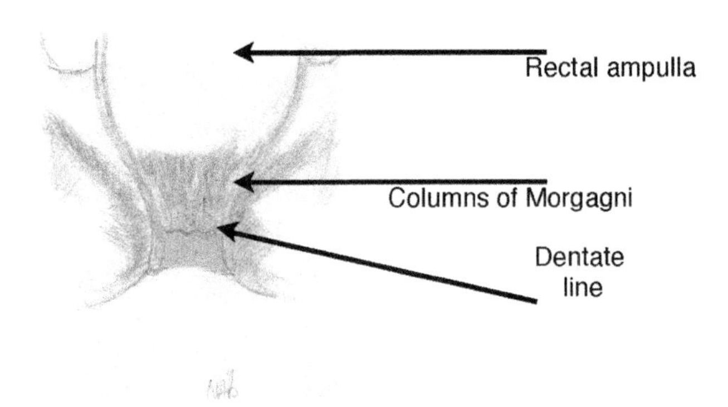

Fig 2.4

This is so named because of a fanciful resemblance to a row of teeth. The author recommends that the alternate, less used name, the Pectinate Line, be discarded, to avoid confusion. It is located approximately 5/8 inches above the anal verge, midway up the anal canal.

The D L is a wavy horizontal landmark dividing the anal canal into distinct upper and lower halves, each with a different lining. Its appearance is due to small valve cusps which span the distal ends of the grooves between the vertically disposed rectal Columns of Morgagni. The cusps create crypts, small pockets in the mucosa, into which drain the Anal Glands.

The D L is of disputed origin. Despite conflicting evidence, the most commonly held and convenient view, is that it is the remnant of the embryonic Proctodeal membrane. This possibly incorrect but simplistic belief avoids making it complicated.

The Mucosal Lining

Above the D L, the lining is a continuation of the columnar or cuboidal mucosa of the colon. Of the same pink color, it is lax and thrown into longitudinal columns (of Morgagni) with intervening grooves, which end at the anal valve cusps. This lining is innervated by autonomic nerves and the mucosa is insensitive to pain. The underlying venous channels, lying within the submucosal space, are tributaries of the Portal System.

Below the D L, the canal is lined with squamous epithelium, identical to perianal skin, but devoid of hair, sweat follicles and sebaceous glands. Here it lacks a well defined submucosal space, is innervated by somatic nerves, and is exquisitely sensitive to painful stimuli. This lining is called the Anoderm. It is pale in color although the underlying venous plexus may impart a blueish tinge. The venous (External Hemorrhoidal) channels belong to the Systemic circulation.

There is a narrow ill-defined whitish area of transitional epithelium between the dentate line and the proximal Anoderm, sometimes referred to as the Pecten Band.

The submucosal and subcutaneous spaces of both halves, though separate, connect loosely, thus allowing a Portal-Systemic connection. The spaces are imperfectly separated by the Suspensory ligament of Parks*. This theoretical structure, which has never been convincingly demonstrated, has been the subject of much discussion and varying opinions. Parks ascribed it to the terminal fibers of the Conjoined Longitudinal muscle, others suggested that it originated within the submucosa of the anal canal. Whether it in fact exists is also moot but what is inescapable is that the the anoderm below the DL is tightly applied to the underlying internal sphincter and that the loose submucous space of higher up has vanished at that level. This has implications in the evolution of haemorrhoids. and sometimes in their treatment.

As with all tissues throughout the body, advancing years will result in attrition of these structures and in concert with the mechanical effects of repeated bowel movements over time will inevitably lead to eventual weakening of its role in restricting mobility of the mucous membrane in that neighborhood.

The Spaces

There are six well recognized spaces and one potential space. All the spaces have the potential for abscess formation.

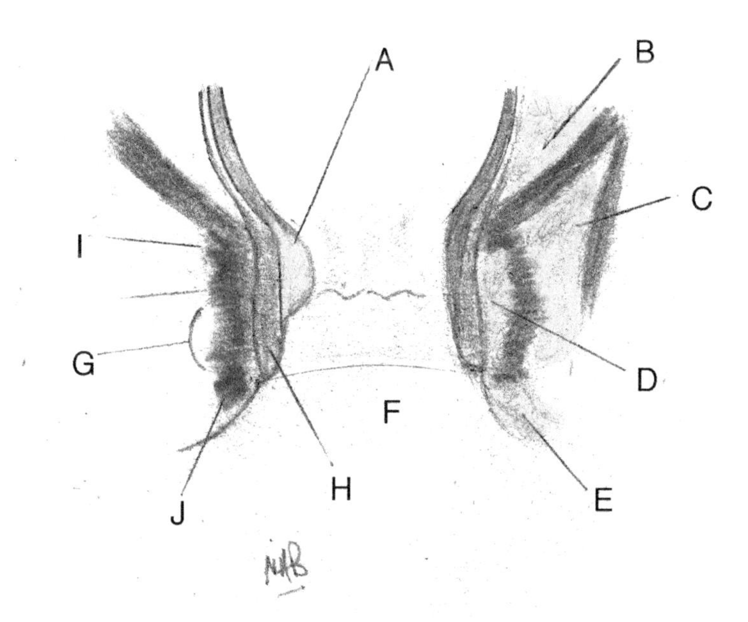

Fig 2.5 A= Submucous B= Supralevator C= Ischiorectal fossa D= Intermuscular E= Perianal F= Postanal G=External Sphincter mass H= Internal sphincter I= Anorectal Ring J= Superficial External Sphincter

1. Ischio Rectal Fossa. C

   (Ischio Anal space in American texts) This is a pyramidal shaped fat filled space lateral to the anal canal,which allows expansion of the bowel to accommodate stool bulk, during defecation. It is bounded medially by the levator ani muscle and the deep external sphincter, laterally by the obturator internus and its fascia and skin superficially

2. Perianal Space or Marginal. E "

   Pedantic texts distinguish between these two spaces but the author believes, because they are so clearly related, that for simplification they should be regarded as one and the same.

   This space contains tributaries and branches of the external haemorrhoidal vessels.

3. Submucous. A

   This lies above the dentate line. It is lax and contains superior haemorrhoidal vessels. Not only is this space a site for abscess formation but the laxity of the mucosa is an important factor in the development of internal hemorrhoids. The diagnosis of submucous space abscess is frequently incorrect: this is most likely a misdiagnosed intermuscular abscess and it is easy to mistake one for the other. This is more academic than important since the treatment is identical viz. drainage into anorectum in each instance.

4. Supralevator. B

   This area is intrapelvic and extraperitoneal. Although it traditionally is regarded as an Anal space it is in reality pararectal in location.

   The intrapelvic intraperitoneal area is beyond the scope of the title of this book. However it is within the realm of the General Surgeon and it may be the site of an abscess secondary to perforated diverticulitis, appendicitis and gynecologic infection and this may rarely track downwards and present as a perineal or ano rectal abscess/ fistula.

5. The Intermuscular (Including the anal glands). D

   This is a potential space between the internal and external sphincters and houses anal glands.

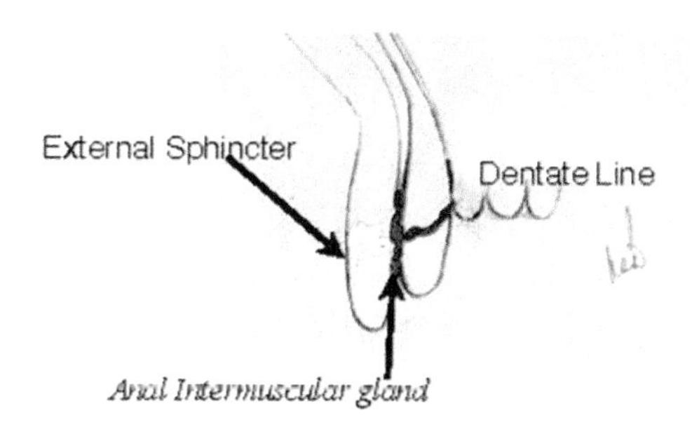

Fig 2.6

These glands are racemose and branch in different directions: up and down, medially and laterally both inter and transmuscularly. They communicate with the anal canal lumen inside a valve cusp at the dentate line. This latter feature is a constant and is of great importance in the surgery of fistula.

In keeping with Eisenhammer's time honored concept, all fistulae and many abscesses start here.

6. The Post Anal Spaces. F

The superficial post anal space is a continuation of the perianal space. The deep space lies deep to the ano coccygeal ligament and lies between the back of the anorectum and the sacrum and coccyx.

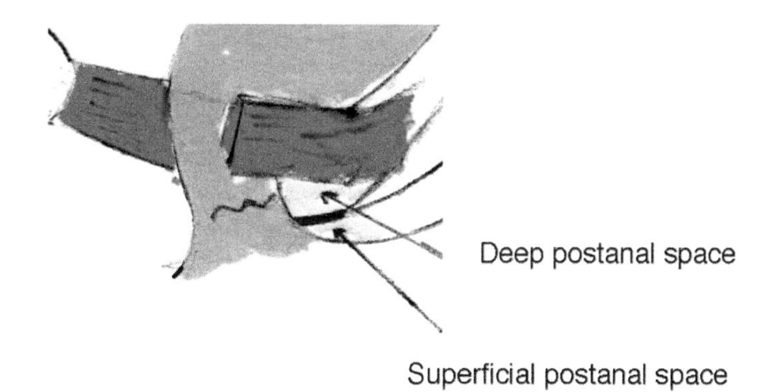

Fig 2.7

Rectal Bleeding

The presence of blood in or with the stool demands evaluation of the entire colon. The color of the blood (dark or light colored) is a pointer towards the likely origin of the bleeding source. Blood originating in the proximal colon will by the time of its exit, have turned darker. Fresh bright colored blood will likely have originated in the distal colon or anorectum. Although this may be influenced by numerous factors such as intestinal hurry, concomitant diarrhea etc, it is a useful guide.

CHAPTER 3

EXAMINATION OF THE ANUS AND RECTUM

Wherever possible, if circumstances permit, before endoscopy or contemplated surgery, prepare the bowel with one or two phosphate (Fleet) enemas.

There are two common positions for the patient; left lateral decubitus and knee elbow. (Genu/ pectoral is an exaggeration of the latter). Although both afford excellent access, the author prefers the first because the patient is more comfortable, less exposed, less embarrassed and therefore more cooperative. The patient can be draped or covered to expose only the area in question.

The examiner sits on a stool so that his/ her head is more or less at the same level as the patient.

Good lighting is essential. Magnifying loupes can be extremely helpful and are strongly recommended.

Gentleness is a prerequisite. If the examination is rough it can be unpleasant for the patient. For full cooperation and to minimise anal orifice reflex contraction, *always prepare the patient by telling him /her what you are doing and what to expect.*

Visually inspect the anal area and adjacent perineum. Instruct the patient is to strain down. This may reproduce prolapse of the hemorrhoids or other pathology.

Examine the area for tags, swellings, local hygiene, condylomata, encrusted fecal matter, leaking stool, mucous, excoriation from scratching.

Press the pulp of a *well lubricated* gloved index finger flat and firmly but gently against the anal orifice. Wait a few moments; the orifice will then relax and the finger can be slipped inside, producing minimal discomfort. Gently examine the canal bidigitally, noting the tone of the sphincters. The palpable depression demarkates the inferior limit of the anal intermuscular space i.e. the space between the internal and external sphincters.

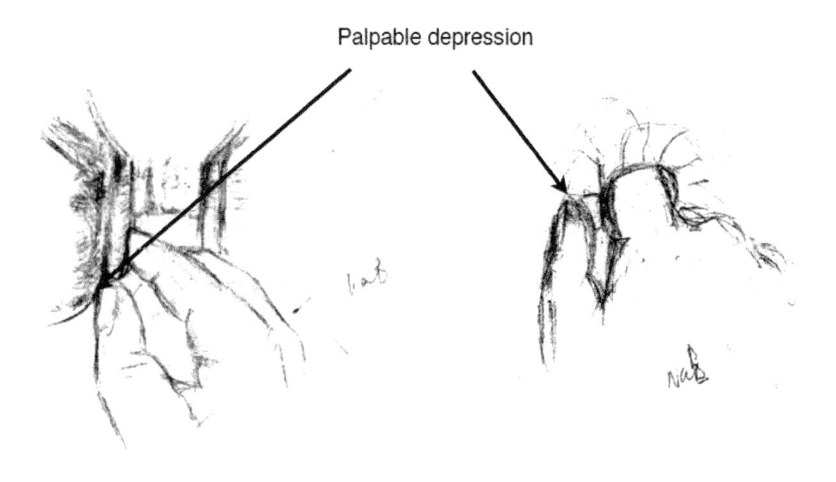

Fig 3.8 Once bidigitally positioned , the examining hand is swivelled to and fro, alternating supination with pronation.

Next, alternately pronate and supinate your hand , thus palpating the depression and internal sphincter circumferentially. This exquisitely imparts subtle information about areas of induration, masses and nodules etc. Then instruct the patient to contract the anal musculature and repeat the bidigital exam while so doing.

Next, gently insert the finger for its full length. The examination should be thoughtful and the local extraluminal structures envisaged mentally. With patient's voluntary contraction, the anorectal ring can be clearly felt posteriorly due to the sweep of the pubo rectalis sling, and the sphincter tone clinically assessed. Constantly remember that the ano rectal ring is approximately one inch only from the anal verge. This is of paramount importance in surgery for fistula. The puborectalis sling is palpable laterally but anteriorly it is deficient.

Anteriorly are located the recto-vesical pouch and the prostate or recto- uterine pouches, the uterus and cervix.

Laterally are the ischium and I/ R Fossa and posteriorly the sacrum and coccyx with the post anal spaces in between them and the bowel.

Carefully feel for intraluminal masses, indurated areas or any other abnormality. On withdrawing the finger it is wiped on toilet tissue for macroscopic evidence of blood.

A proctoscope (a.k.a anoscope in American texts), well lubricated and moderately warmed (NOT HOT) is then gently introduced, first aimed forwards toward to pubis and then angling it backwards to follow the ano-rectal angle. Remove the obturator. Cotton wool pledgets on forceps should be available to clear the path of stool. Instruct the patient to strain down upon the inserted proctoscope. This will reproduce prolapse of the mucosa including internal hemorrhoids (which may show some bleeding) , polyps into the lumen of the instrument. Slowly withdraw the scope whilst the patient is straining down.

Rigid Proctosigmoidoscopy

Prior to the era of fiberoptic flexible colonoscopes and sigmoidoscopes, this was the last word in evaluation of the rectum and colon. Nonetheless, there is today still a role for it as an easily performed office procedure, a first step in evaluating the rectum and distal colon.

Some advocate not preparing the bowel. I do not support this practice for several reasons. It is in the first place more aesthetic to examine a clean bowel. Secondly it affords an excellent circumferential intraluminal view without visual obstructions or decoys. It is a fallacy that cleansing the anorectum will remove evidence of bleeding or confuse its source. On the contrary, so doing will ensure a more reliable demonstration and identification of problem areas.2q

First clear the distal bowel by one or two or, if necessary, even three, phosphate (Fleet) enemas. Gentleness is paramount; the examination should not be painful. The lumen must be constantly visualized and never lost sight of. Exercise care at all times, and especially where there has been previous lower abdominal surgery. Keeping a mental picture of the natural curvatures of the ano rectum (two to the right and one to the left) , including the three rectal valves (of Houston), the well lubricated, warmed scope can usually be introduced with no or minimal discomfort. Keep in mind that the middle valve indicates the level of the peritoneal reflexion. A perforation above this will be intraperitoneal and conversely, below, will be extraperitoneal and into the supralevator space.

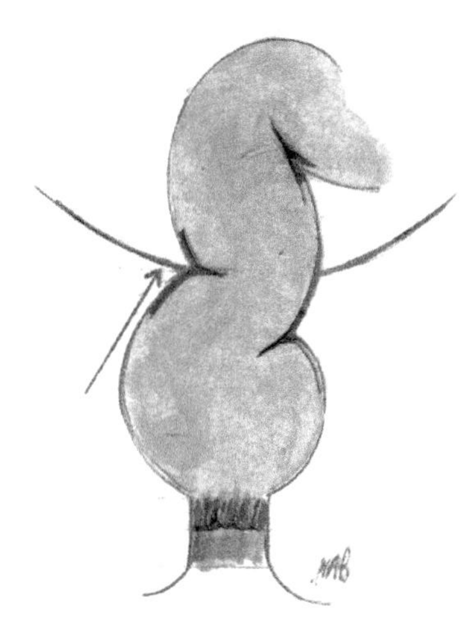

Fig 3.9 The middle rectal valve (of Huston) indicates the level of reflexion of the parietal peritoneum (PP).

Suction should be constantly available to help clear the path. Use the bellows minimally, just a few puffs only to gently separate the walls for visibility; do not pump up the bowel. In the event of untoward pain, withdraw the scope partially, pause and then proceed with great caution. Be prepared to abort the procedure if you deem it dangerous to persist.

There are many models of rigid Sigmoidoscopes, both reusable and disposable. The usual length is 25 (10 inches) cms but 35 cm lengths are available. Proximal lighting , reflected off the well polished and shiny interior walls of the scope, ensures continual good visibility.

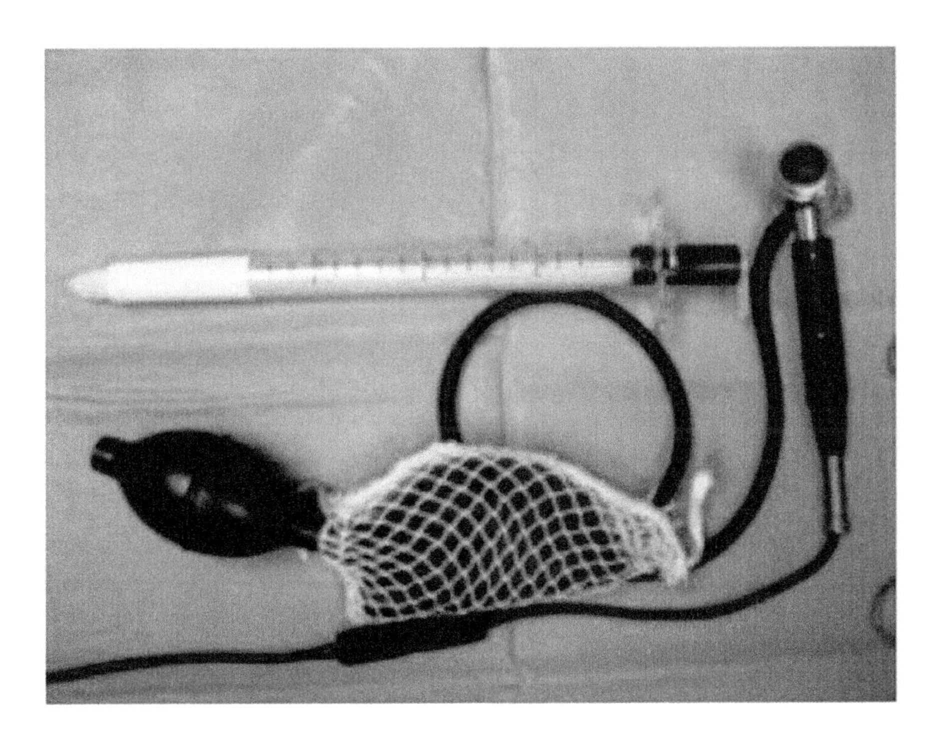

Fig 3.10 Disposable rigid Sigmoidoscope

Procedure

This can be performed in the left lateral position, the knee/ chest position, or when at the time of surgery, in lithotomy. The author's preference is the left lateral position with the buttocks projecting slightly over the side of the examination table. A small supporting cushion may be helpful. You may find it more comfortable to be seated, providing that your eye is positioned at the correct level.

Gentleness is paramount. The scope is warmed (not hot!) and well lubricated. The examination should be unhurried.

The lumen is constantly visualised as the scope is advanced.

The natural curves are followed.

The bellows are minimally used; only an occasional puff to separate the walls and keep the lumen in the view.

Biopsies are taken as needed. Above the dentate line this is painless.

It is the rule that the scope can be easily inserted for its full length. An additional small puff of air at this level, will open up a further length of the proximal bowel, often extending visibility for a significant distance beyond the end of the scope.

Pain is a warning signal. If it occurs, withdraw the scope for a few inches and after the pain resolves, you can again gently advance.

The bowel is again carefully viewed circumferentially during withdrawal, all the while gently wiggling the scope. If conscientiously viewed, there should be no blind areas.

Flexible Sigmodoscopy/ Colonoscopy

This procedure, vastly superior to rigid endoscopy, allows much more extensive and reliable evaluation of the rectosigmoid and left colon. The downside is that the biopsies are of much smaller size than possible through the rigid scope.

The bowel is prepared with several (up to three) Fleet enemas.

You can easily and comfortably carry out the examination seated, a great boon for your back!

Modern scopes, containing a camera in the tip, are vastly superior to early fiber optic models which had an eyepiece for direct viewing. The now greatly magnified image on the monitor screen, provides the operator great freedom of movement and the old problems of broken fibers which interfere with the image, are absent. The scopes are more flexible than the older ones , hence safer to use, and are also less fragile. They must nevertheless at all times be treated gently since repairs and maintenance, apart from the inconvenience of unavailability of equipment, are extremely expensive.

The patient is positioned in left lateral decubitus. Sedation should not be needed. Gentleness and patience are essential and there should be minimal to no discomfort.

It is essential to familiarise yourself with the controls, channels and capabilities of this equipment before setting out to use it. You require no specialised training once you are familiar with the controls and the channels.

There are several excellent brands of scope on the market which for decades has been monopolised by Japanese optical companies. The insertion of the scope is similar (minus obturator) to the rigid type and the same principles apply. You must see the lumen at all times and not advance the scope nor vigorously manipulate the tip unless the lumen can be seen, since there is a risk, albeit small, of perforating the bowel, especially at flexures and curvatures. Again, don't over insufflate.

Endoscopic skills are rapidly acquired and most times the entire left colon is visualised, up to at least 80 to 90 cms from the anal verge (close to or at the splenic flexure in the average adult patient). Although total colonoscopy is beyond the scope of this book (no pun intended), for those surgeons who are qualified to perform colonoscopy, as a bonus, it is often possible to navigate the rest of the colon, which may, surprisingly, be totally clean and perfectly prepared from the enemas only, without the need for any additional oral preparations.

Position for Surgery

Unless otherwise indicated all operative procedures described are performed in the Lithotomy position with the table in varying degrees of Trendelenburg tilt, as needed. It is helpful also to raise the table so that the field is comfortably at the operator's eye level.

Anal Stretch

This is a prelude to many of the procedures described in this book.

The patient must be fully relaxed and under General Anesthesia. A Lord's stretch (massive stretching plus anodermal tearing) , once quite popular for hemorrhoids, is not performed. It is traumatic to the sphincter muscles, it's benefits questionable.

The stretch employed in this work serves two purposes:

1. It reduces, minimises or even circumvents post operative pain and;
2. It facilitates the atraumatic insertion into the anal canal of instruments including specula. It is NOT a sphincter tearing procedure and its effects are intended to last for no more than a week or two.

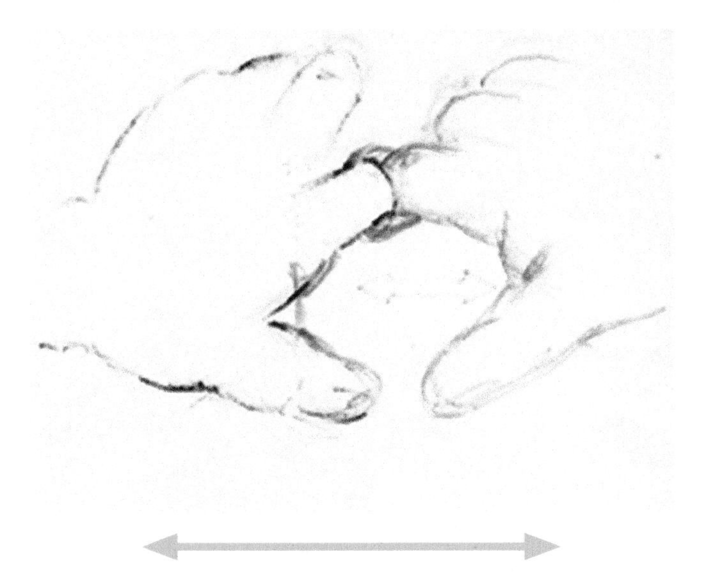

Fig 3.11 Stretch strictly horizontally

The anal canal is slowly stretched to a maximum of four fingers. The action is slow and deliberate and "ripping" is to be strictly avoided. There should be no tearing and ideally, no or minimal bleeding.

The correct direction is strictly horizontally, where the ischial tuberosities restrict the possibility of over stretching. Diagonally directed stretching is to be avoided.

Personal communication. The practice of this adjunct to excisional haemorrhoidectomy was introduced to the author by the late B T le Roux, former Head of Cardio/ Thoracic Surgery at Wentworth Hospital, Durban, S. Africa, when, in 1953, the author was a surgical resident at the Royal Infirmary, Edinburgh, Scotland and Ben was Registrar in General Surgery. Preliminary stretching will diminish postoperative pain significantly, even eliminate it completely. However, there are occasions however when it may be contraindicated: see comment on page 33.

CHAPTER 4

HAEMORRHOIDS ("PILES")

While constipation and excessive straining at stool without question contribute, hemorrhoids frequently develop in their absence. The majority of the population over the age of fifty will demonstrate hemorrhoids, even if asymptomatic, on examination. In the absence of symptoms, they require no treatment.

Pathology

Hemorrhoids are often described as "varicose veins" of the anus. Although a colorful description, this is simplistic.

Internal haemorrhoids and external haemorrhoids, although they frequently co- exist, are different entities. The venous drainage of the internals is mainly into the Portal Circulation whereas in the externals it is Systemic. Thus with portal hypertension, varices may occur in the anal canal similar to those in the oesophagus. This, though often described in textbooks, is largely theoretical and in practice is very rarely seen.

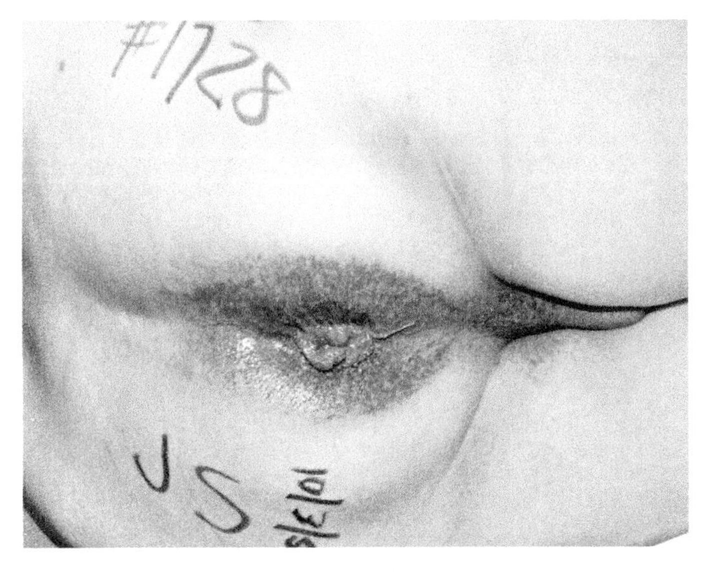

Fig 4.1 Minimal External haemorrhoids

The anatomic location of the three primary internal haemorrhoids is constant at 3 (left lateral), 7 (right posterior) and 11 (right anterior) o'clock, as viewed in the lithotomy position. Purists have attempted to provide an anatomic explanation for this but the evidence remains unconvincing.

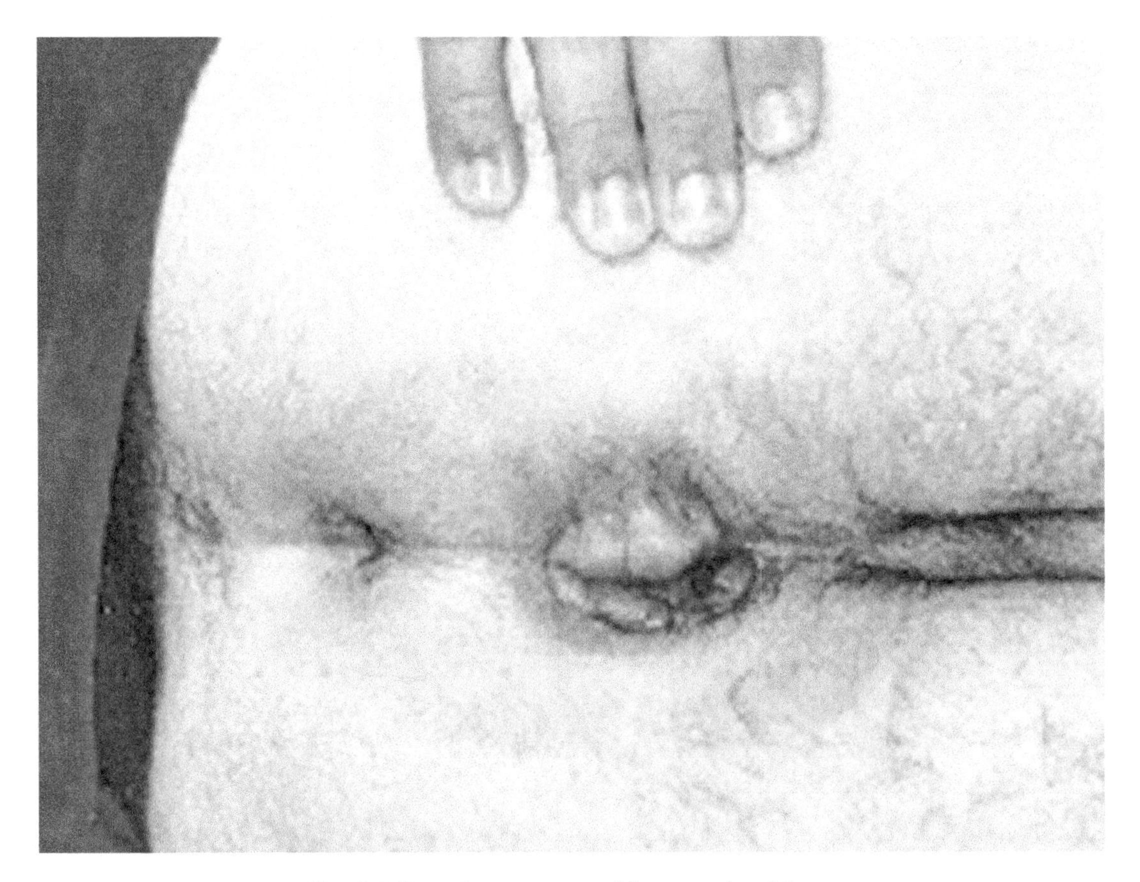

Fig 4.2 Significant external haemorrhoids/ tags

Internal hemorrhoids are in fact an exaggeration of the normal local anatomy and constitute a varying degree of prolapse of the ano rectal mucosa. With excessively lax mucosa and a rich vascular content, both arterial and venous, they are a product of years of repeated mechanical propulsive forces in concert with degeneration and attrition of local supportive tissues.

Internal Haemorrhoids

While these are more significant diseases with more serious manifestations such as chronic blood loss, irreducibility and strangulation, the external variety are for the most part more symptomatic and therefore more undesirable in the patient's eyes. As the internal haemorrhoid mass evolves, the dentate line later comes to lie on its lateral side.

This is due to the relative fixation to the underlying internal sphincter of the anoderm below the dentate line and the "folding over" of the highly mobile supradentate mucosa as it descends. During defecation the rectum is unable to distinguish between intraluminal stool and mucosal masses and attempts to expel both. Parks postulated the presence of a "Mucosal Suspensory Ligament" which, when attenuated, eventually loses its ability to anchor the anal mucosa to the underlying internal sphincter muscle at approximately the level of the dentate line.

Internal haemorrhoids are conveniently classified according to four “Degrees” (In American texts, “Grades”). These are not different diseases but manifestations of the severity of the same condition. Staging is convenient for evaluating and documenting.

What may have at one time been a first degree hemorrhoid will in time inexorably progress to and beyond the next stage.

First Degree: Bleeding without external prolapsed

Second degree: Bleeding plus prolapse which self reduces.

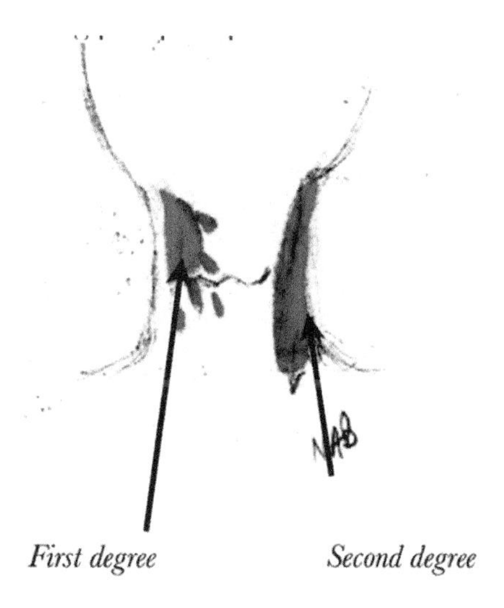

Fig. 4.3

Third degree: Prolapse requiring digital replacement.

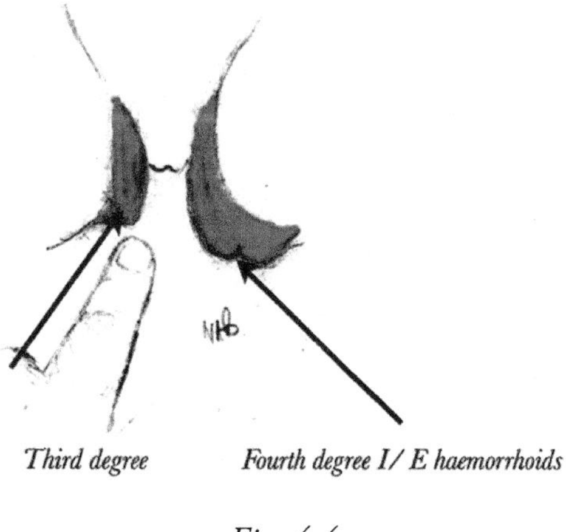

Fig. 4.4

Fourth degree: Irreducible permanent prolapse, unrelated to passage of stool.

Mucous leak from higher up can be a troublesome problem. Some texts further subclassify into a fifth degree. I believe that this creates unnecessary confusion.

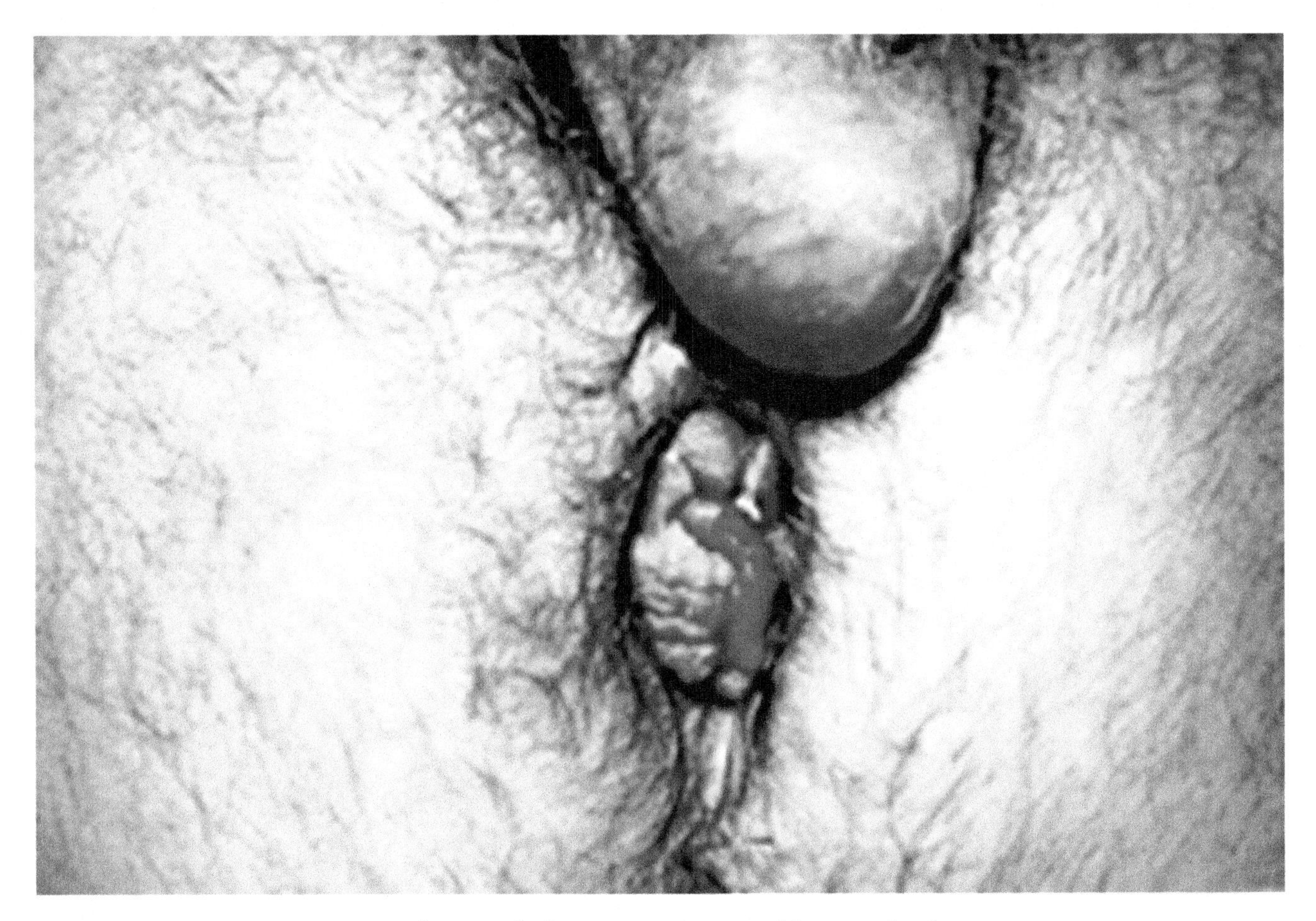

Fig 4.5 Fourth degree intero/ external haemorrhoids .

Complicated Haemorrhoids

These are rarely seen in developed countries but are commonplace in the third world. Prolapsing internal hemorrhoids may "become stuck" because of sphincter spasm, which in turn "traps" the prolapsing tissues. Venous occlusion rapidly leads to severe edema and this in turn may secondarily compromise the blood supply. These may then be called Strangulated Hemorrhoids. Edema may extend to the perianal area. Patchy gangrene may ensue with secondary infection.

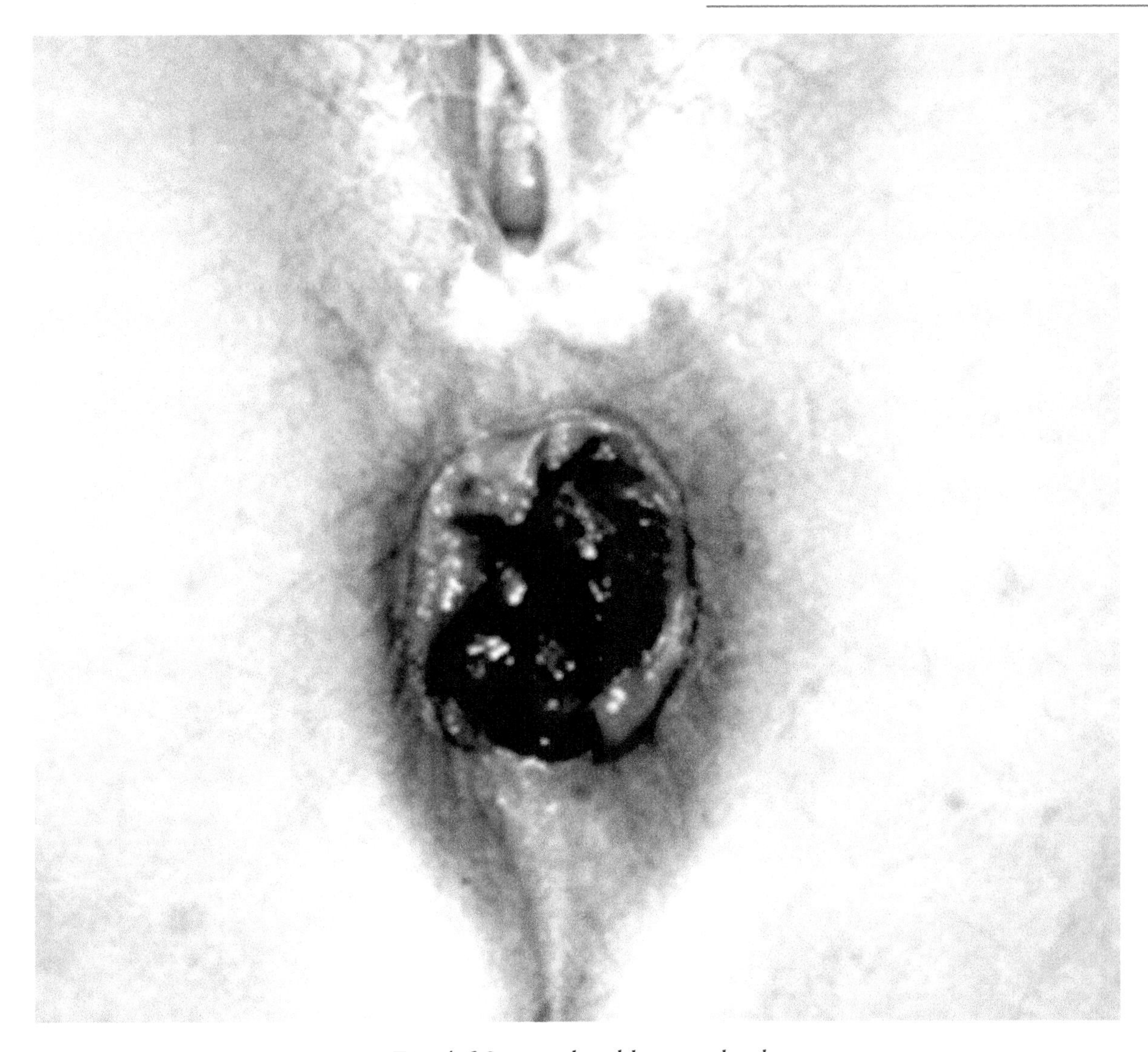

Fig. 4.6 Strangulated haemorrhoids

Such cases will mandate surgery but its timing is controversial. Because of the fear of provoking local cellulitis and septicemia and also the distortion of local anatomy from the swelling, the time honoured approach has been to initially treat the area conservatively to achieve local resolution and then subsequently perform elective hemorrhoidectomy when local conditions were more conducive.

The addition of the gentle local stretch at the onset of surgery will often magically create optimum conditions allowing one to then proceed with excision. If not, a stretch, part one of a two stage procedure, will significantly speed the process along.

<u>Comment</u>

A word of caution: If there is severe secondary infection present, stretching is to be avoided. Under these circumstances it is safer to allow local conditions to settle with conservative treatment before proceeding. Do not risk producing an uncontrollable infection, possibly Fourniers gangrene, antibiotic availability notwithstanding, by rushing in to perform surgery inadvisedly early to gain one or two days. Here discretion is truly the better part of valor.

Conservative Treatment

Elevate the foot of the bed to a comfortable angle.

Apply local dressings of hypertonic saline combined with ice packs. Refresh every four to six hours.

Prescribe stool softeners.

Resolution is rapid, usually two or three days.

Symptoms

The cardinal symptom of internal hemorrhoids is bleeding at stool. This blood may be dark (venous) or bright red (arterial) or a mixture of both. Quantities may vary from just staining the toilet paper to a significant visible bleed into the commode. It is also not uncommon because of long periods of hardly noticed, continual minor bleeding, for patients to present with iron deficiency anemia.

Prolapse occurs in the later degrees.

Treatment

The objective of all forms of treatment is twofold:

1. Removal of the hypertrophic mass of the pile and
2. Restricting mucosal mobility. This is achieved by promoting adherence of anoderm to the surface of the underlying internal sphincter, even obliterating the submucosal space in some areas.

Options

Asymptomatic internal hemorrhoids require no treatment. Symptomatic first and second degree piles can be treated by simple office based methods such as injections (sclerotherapy), application of rubber bands, infra red or cryotherapy. Excision should be reserved for third and fourth degree piles.

Note: Internal haemorrhoids can exist without external components and vice versa. Operative techniques may have to be tailored according to local circumstances.

External haemorrhoids are situated in the marginal area. Although their venous component is in the systemic circulation, they frequently coexist and anastomose with the portal e.g. with third and fourth degree internal haemorrhoids, They may be fibrosed and are then commonly known as "External tags". These are far more prone to be symptomatic than their internal counterparts, and the symptoms may even be severe and debilitating. Their skin is exquisitely sensitive to pain. Particles of stool may get caught up as may be mucous, leaking down from above. It can be difficult to maintain local hygiene and the area may be inflamed or excoriated from chronic contact dermatitis and/or scratching. Wherever formal haemorrhoidectomy is carried out, it must include both internal and external components. Even in the absence of significant internal hemorrhoids, externals may merit excision in their own right.

Complicated by hematoma or thrombosis, an external haemorrhoid can be one of the most painful anal lesions. It is sometimes described as a "two week self curing lesion". Unfortunately, if left to self cure, it will likely leave a troublesome skin tag. Occasionally they may ulcerate and then continue to bleed. Surgery, usually an office procedure, affords instant relief.

Sclerotherapy for 1st and 2nd degree internal haemorrhoids:

Five percent phenol in vegetable (almond) oil, a time honored standby, is a sclerosant designed specifically for internal haemorrhoids. There are however a variety of sclerosants available.

The patient is positioned in left lateral decubitus with the knees drawn up and the buttocks slightly over the edge if the table. The operator is comfortably seated on a mobile stool. Good lighting is essential. The procedure should be painless. The old fashioned Gabriel reusable syringe and needle are obsolete. A 10 cc disposable Luer lock B D syringe with proximal lugs and a fine (#20) spinal needle are ideal.

Since the fluid is thick and does not flow easily through the fine spinal needle, the syringe is filled from the top and the plunger inserted afterwards. The sclerosant is first gently warmed to improve its flow, by immersing the ampoule in a dish of hot water for a minute or two, before loading the syringe.

Fig 4.7 Filling the syringe with sclerosant

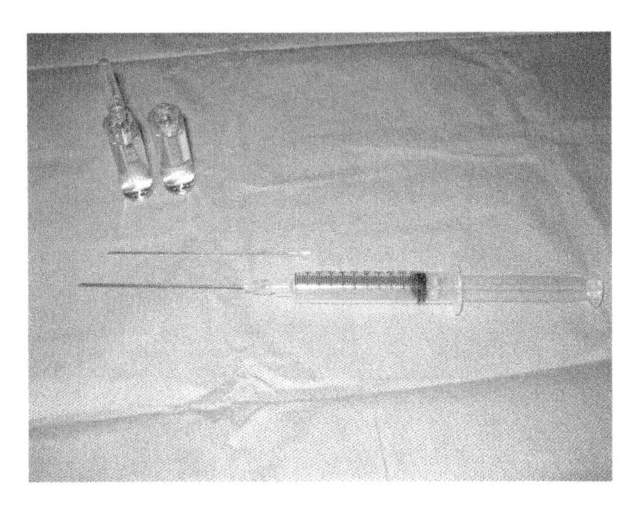

Fig 4.8 Syringe, spinal needle attached and sclerosant

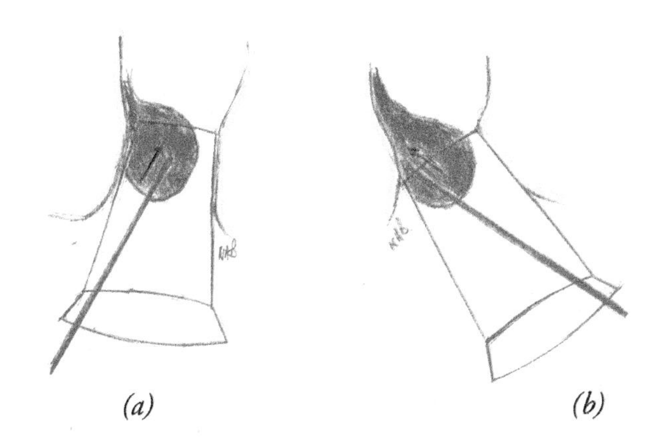

Fig 4.9 As depicted in (a) the procedure is safe. Avoid as in (b) where there is a risk of penetrating the bowl wall

Insert a well lubricated proctoscope for its full depth. Next, whilst the proctoscope is being slowly withdrawn, the patient is instructed to strain down. This forces the internal haemorrhoids, which are recognized by their plum red color, to descend into the scope. Identification is also facilitated by recognising the Morgani columns. The proctoscope is then held in position and the needle is easily inserted into the area of the pedicle, at the upper reaches of the columns, above the DL.

You must exercise the utmost care to not penetrate the full thickness of the bowel wall and especially, in the area of the right anterior haemorrhoid , not to penetrate the prostate gland. To avoid this, hold the needle parallel to the wall of the proctoscope and slowly and deliberately pierce the mucosa. Do not jab. As in the above exaggerated illustration Fig 4.9 , placing the needle as in (a) is safe. Placing it as in (b) must be avoided since if it is advanced it may penetrate the bowel wall and beyond. It cannot be sufficiently emphasized that the sclerosant must be placed submucosally and extravascularly.

The mucosa swells as the space is distended and it turns pale. The appearance has been likened to the skin of a frog's belly. If it blanches early on, the needle point must be repositioned. This procedure, which takes about one minute, should be painless. Three to five ccs are injected into each haemorrhoid mass; one only per session, since after the placement of the sclerosant you will likely obscure the location of the others.

Occasionally there may be negligible bleeding together with some leakage of the sclerosant through the puncture site. This is easily controlled with local pressure from a cotton wool pledget for a minute or two.

The patient should return at intervals of three to four weeks for further injections. If the injections have been well placed, you will probably find that after three treatments, the submucous space has been largely obliterated and further injections do not run in easily.

Treatment will then no longer be needed. However there may be "recurrence" of the haemorrhoids with bleeding after a variable time and this treatment can be safely repeated at a future date, usually years.

Since haemorrhoids and carcinoma may coexist, it cannot be over emphasized that *RECTAL BLEEDING MANDATES EXAMINATION OF THE RECTUM AND COLON HIGHER UP FOR POLYPS AND CARCINOMA, EVEN IN THE PRESENCE OF DEMONSTRATED BLEEDING HEMORRHOIDS.*

Rubber Band Ligation

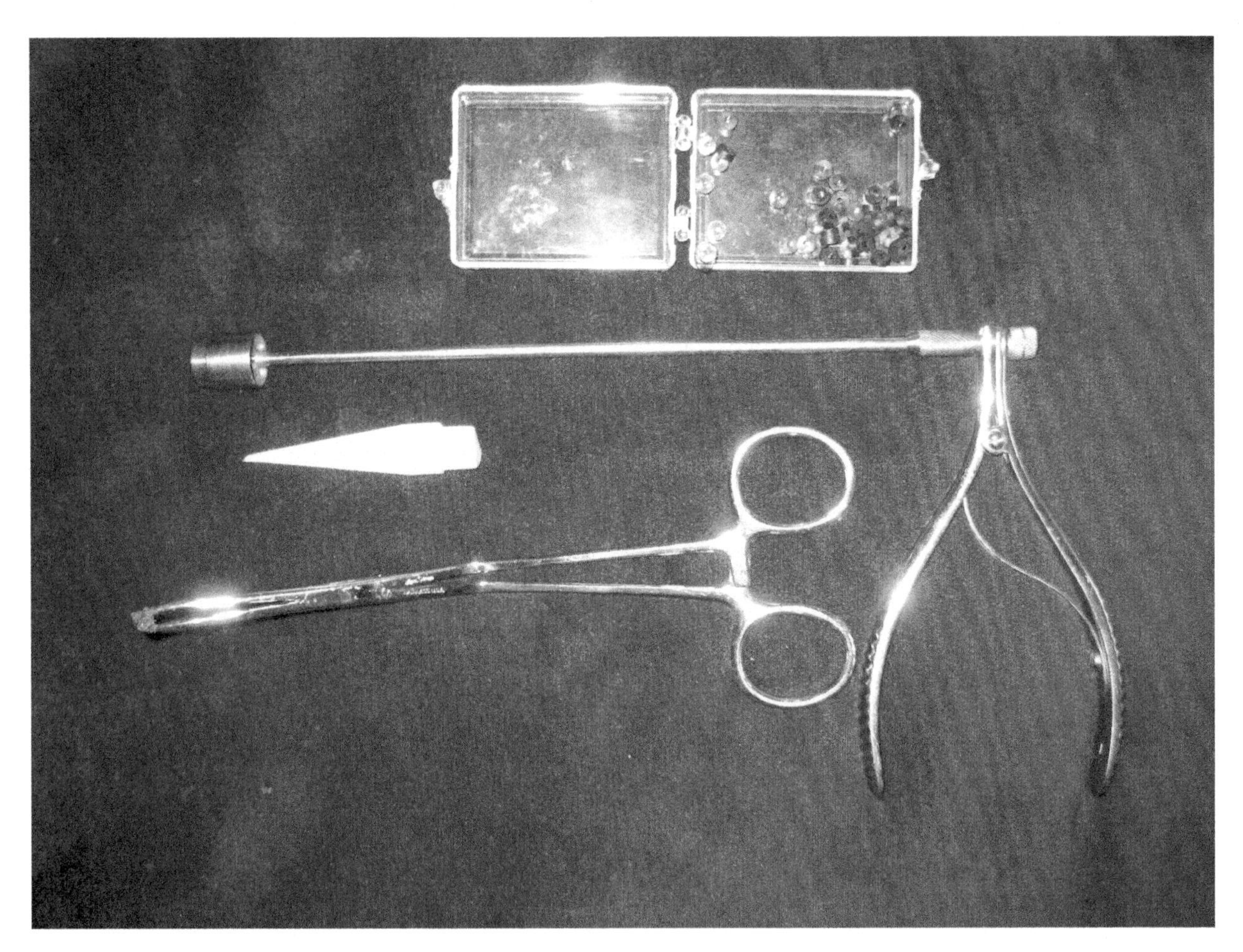

Fig 4 .10 McGivney ligator witrh cone, rubber o- rings and grasping forceps.

The patient is positioned as for injection. You, the operator, are best seated. Good lighting is essential. The McGivney (Fig 4.10) or similar ligator (Fig 4.11) is loaded with an O Ring (Fig 4.12).

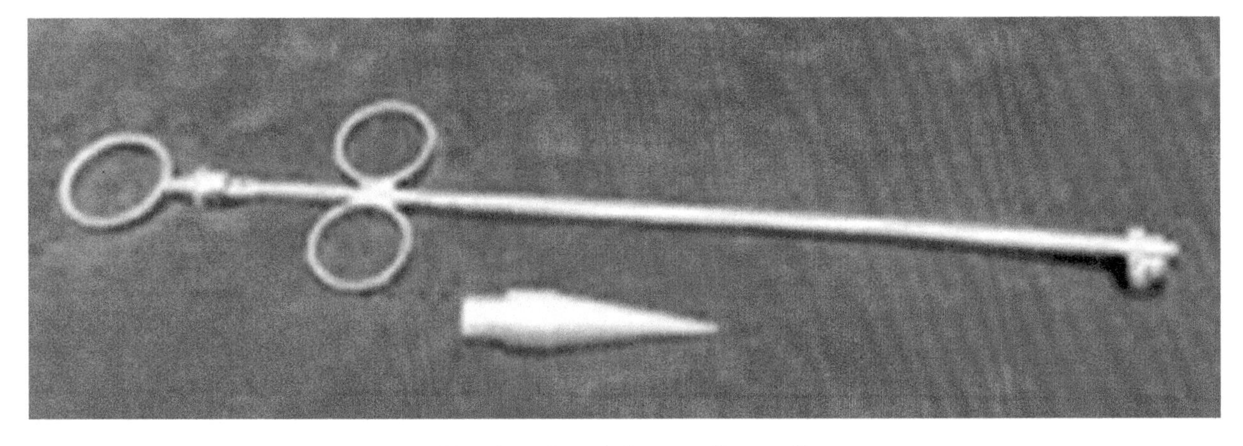

Fig 4.11 Simple ligator (Seward)

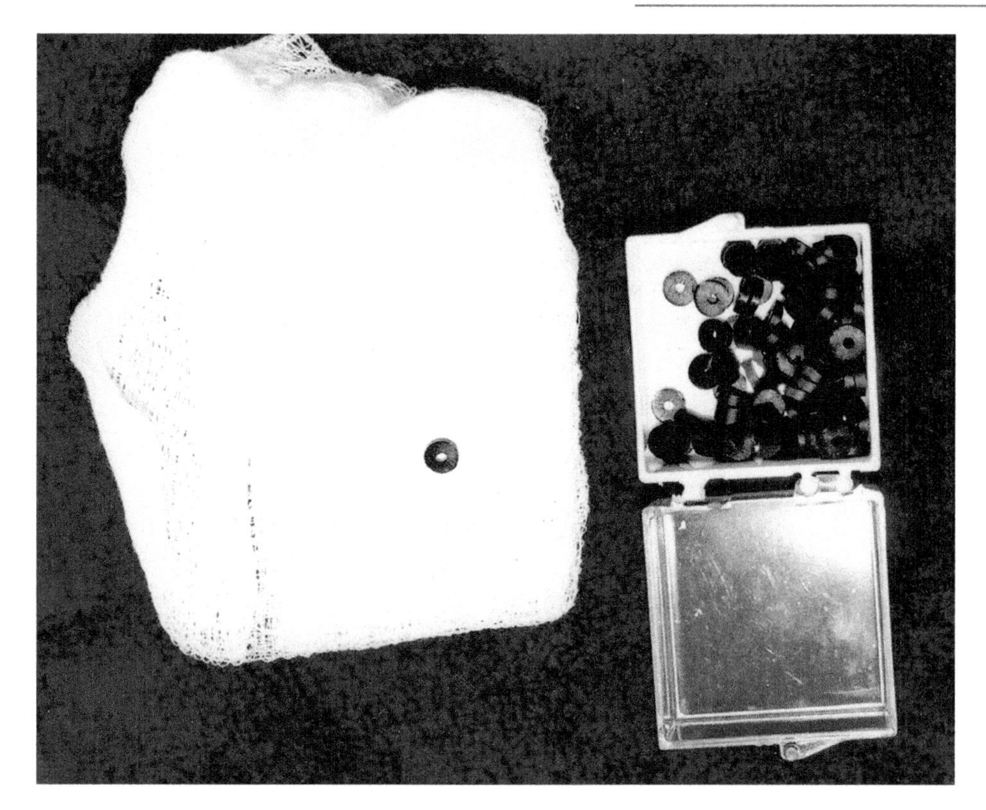

Fig 4.12 Rubber bands. (O rings)

The minute rubber O rings or bands are pushed over the cone onto the drum component of the ligator. The cone is then removed and set aside.

Next, as per illustrations, the hemorrhoid mass is either pulled down into the lumen of the drum or made to descend by getting the patient to strain down. The band is then released onto the hemorrhoid.

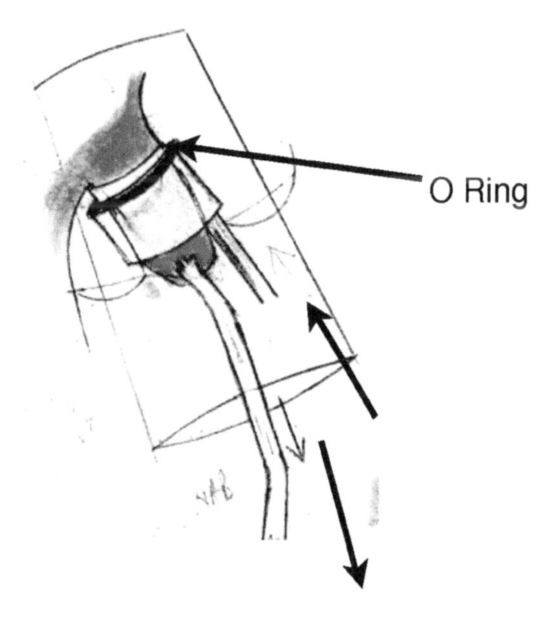

Fig 4.13

After release, the band will immediately contract to its original size, thus securely ligating the tissue contained within the drum. This tissue mass is immediately deprived of its blood supply and venous return and after initial swelling, necroses and shrivels away.

There are two ways of applying the bands.

(a) The pile is viewed through a proctoscope, as described earlier. Then grab the hemorrhoid with the special grasper , passed through the drum of the instrument and pull the hemorrhoid mass down into the drum. The rubber band is then released. This method requires an assistant to hold the proctoscope in position.

(b) An alternative way, requiring no assistant, is to have the patient strain down and simultaneously gently push the drum over the prolapsing mucosal mass and then release the band. This method while effective, takes smaller "bites" of tissue and may therefore entail more visits and treatments than the former.

Remember: *At all times stay above the Dentate line, no matter which method you use.*

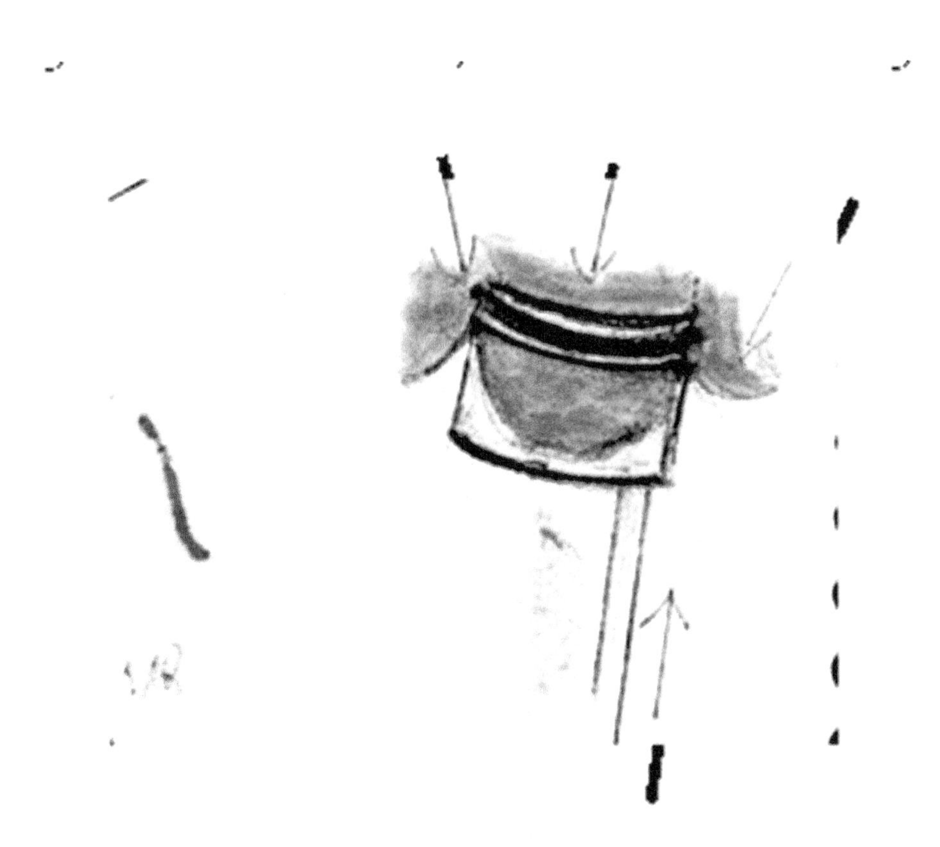

Fig 4.14 Alternative method, without clamp and with patient straining down.

This procedure should produce minimal or no pain. If the patient complains, you have gone too low and inadvertently caught some sensitive up anoderm below the D L. In that circumstance, immediately remove the band; to try and find it a few hours later when it is lost inside an edematous black mass, can be frustrating for the operator and very uncomfortable for the patient.

If you so decide, you can place one or two additional bands. Be sure to leave adequate amounts of mucosa between multiple bands.

The patient returns in four weeks for evaluation and further banding, if needed.

Alternative Methods

The popularity of cryotherapy has faded. It involves expensive equipment with costly maintenance and the author found it to be time consuming and unpredictable. Reports indicated slow and prolonged healing with discomfort and troublesome drainage for possibly months.

Infrared coagulation therapy has been strongly promoted by commercial vendors. The author has no personal experience with it and from all accounts it is costly, and does not appear to offer any advantages over traditional methods.

Hemorrhoidectomy

Open hemorrhoidectomy a.k.a Ligature and Excision method or St Mark's hemorrhoidectomy. This was refined and popularised by C Naunton Morgan and E T C Milligan some seventy years ago. It is rapidly performed and the entire dissection is performed extra anally, affording maximal exposure with minimal instrumentation. Although it is not popular in the United States, it is probably the most widely performed method worldwide, certainly in the U K, the former empire and throughout Europe.

The procedure described is based on the age old method of Milligan and Morgan with the author's modifications, where the ligatures are of non absorbable material and there is no transfixion of tissues. This will ensure that the ties will later be extruded and discarded spontaneously. (The patient must be so informed, alerted and reassured).

Two or three Fleet enemas are administered shortly before giving premedication. This ensures against soiling and it also enables the performance of endoscopy higher up, if needed. "

The patient is positioned in lithotomy. The table is then placed in 20 degrees of Trendelenburg tilt (Fig 4.15). You, the surgeon, are seated and the table is raised so that the anus is approximately at your eye level or where you are most comfortable.

In male patients, suction tubing hooked under the scrotum (Fig 4.16) and trapped within the flexion crease of the groins, secures, lifts and protects the genitalia.

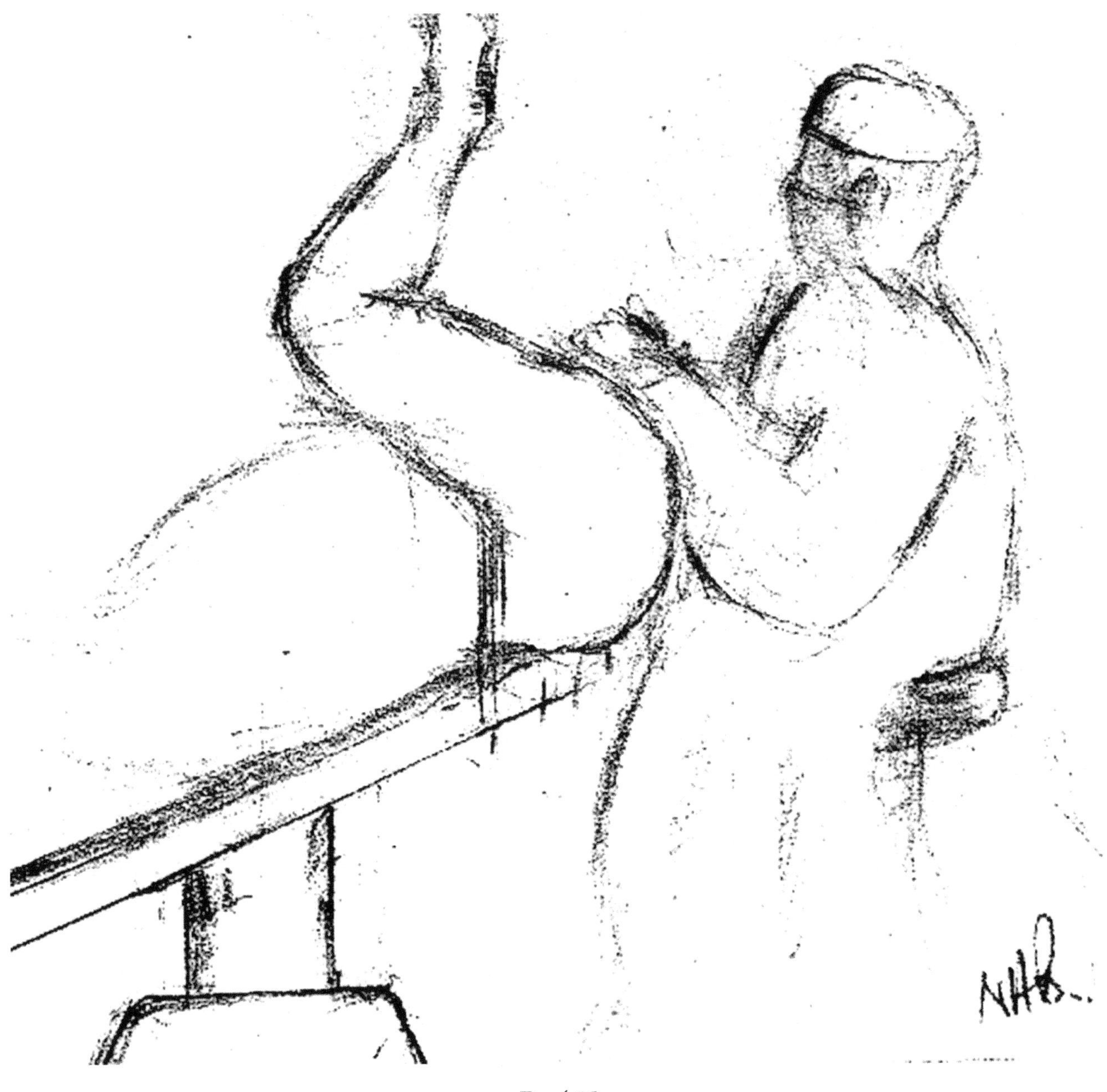

Fig.4.15

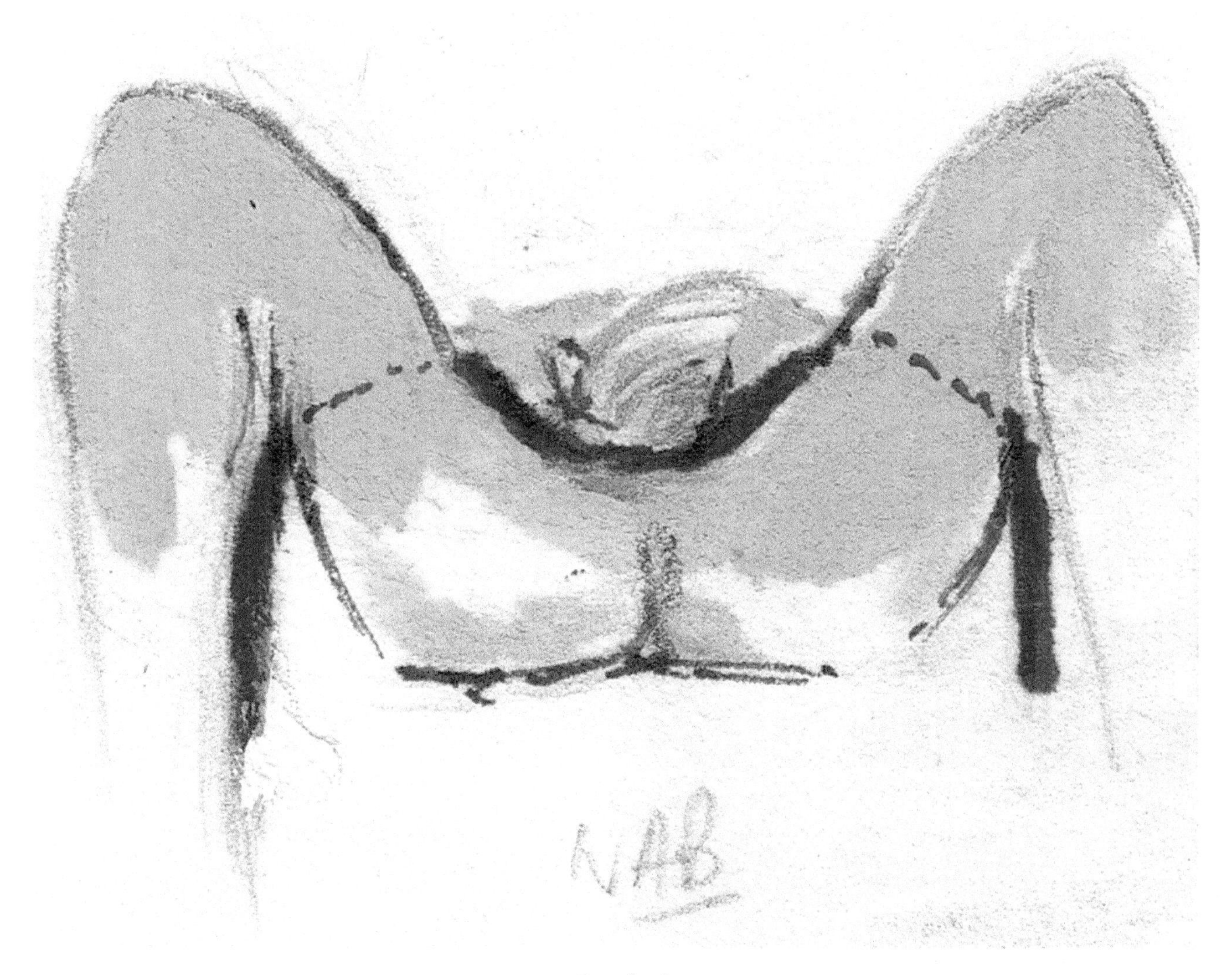

Fig. 4.16

If it has not already been performed preoperatively, endoscopy, rigid or flexible, is now carried out.

It is the author's firm conviction that clean, disciplined steps will inculcate a commitment to the practice of tidy operative surgery. Although the anal area is often regarded as "dirty", all patients undergo a vigorous local prep, just as in abdominal surgery. After shaving, if needed, prepping and draping, inject three to five ccs of 1% long acting lidocaine in 1:100000 epinephrine subcutaneously (extra anally) into each at three, seven and eleven o' clock positions.

Massage locally to disperse.

Fig.4.17

Next, carry out a GENTLE, horizontal four finger stretch of the anus. The fingers are gradually inserted, one at a time. Ideally, THERE MUST BE NO TEARING OF TISSUES.

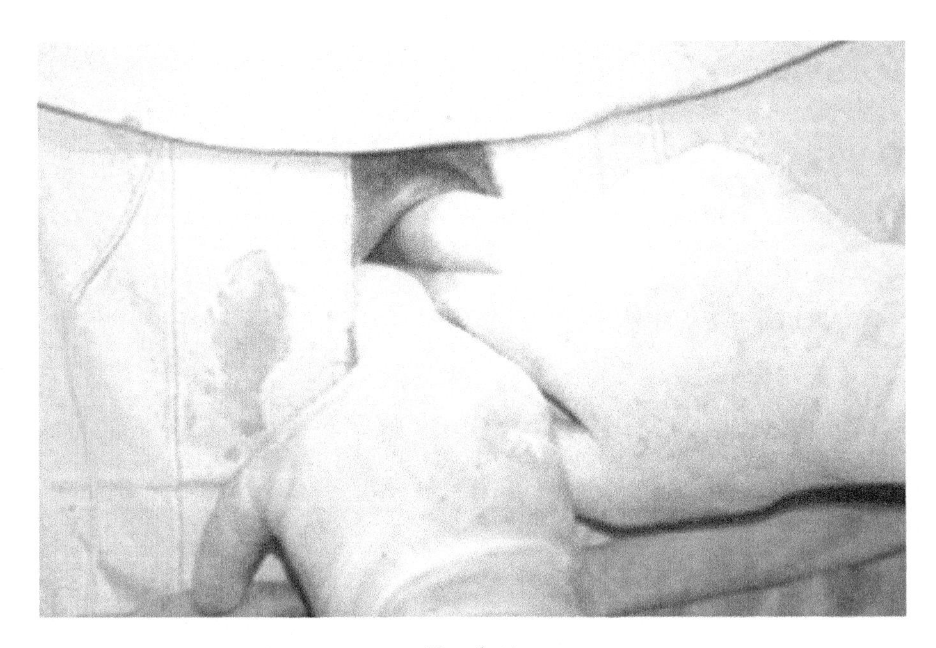

Fig 4.18

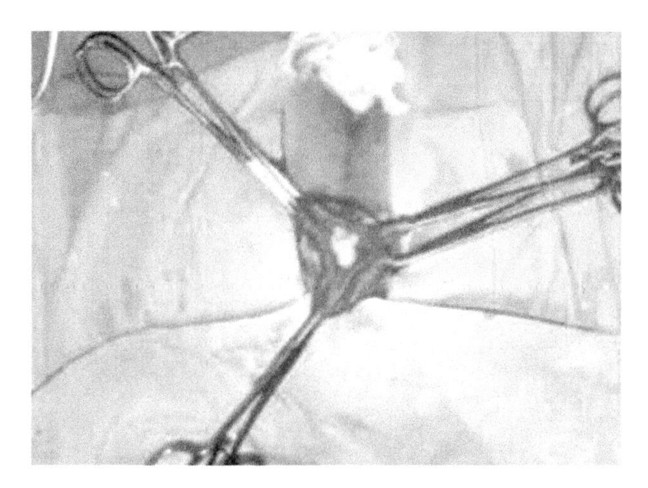

Fig 4.19

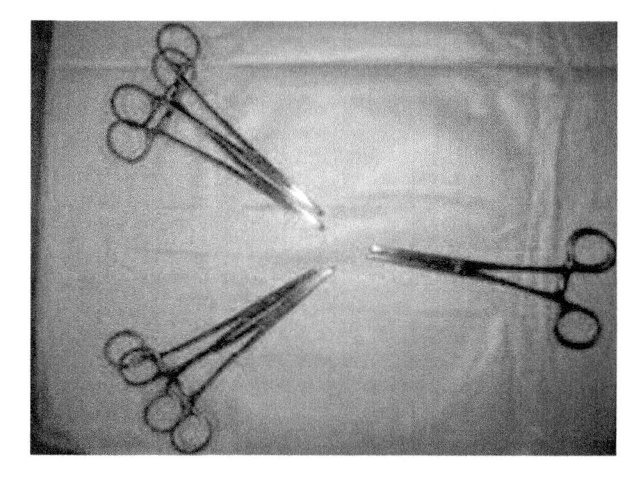

Fig 4.20

This is conveniently named the "Triangle of Exposure"

This way you will ensure that the sphincters are not overstretched and there will be no postoperative incontinence. With the stretching, the raised wheals from the injected local anesthetic disappear as the solution diffuses circumferentially around the anus.

The anal dilatation combined with the local anesthetic will render the patient pain free for the first few postoperative hours and thereafter render the pain tolerable for one to two weeks, during which it easily controlled by mild analgaesia such as oral hydrocodone. Codein is constipating and the patient should be advised to avoid over use.

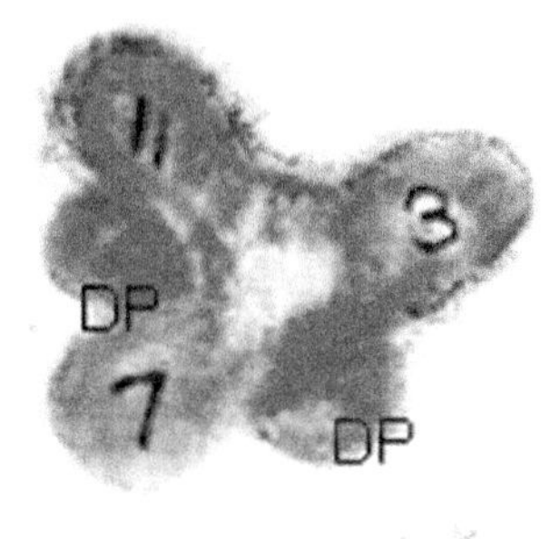

Fig 4.21 Showing two daughter (secondary) piles marked DP. The main three haemorrhoids are in the 3, 7 and 11 o'clock positions

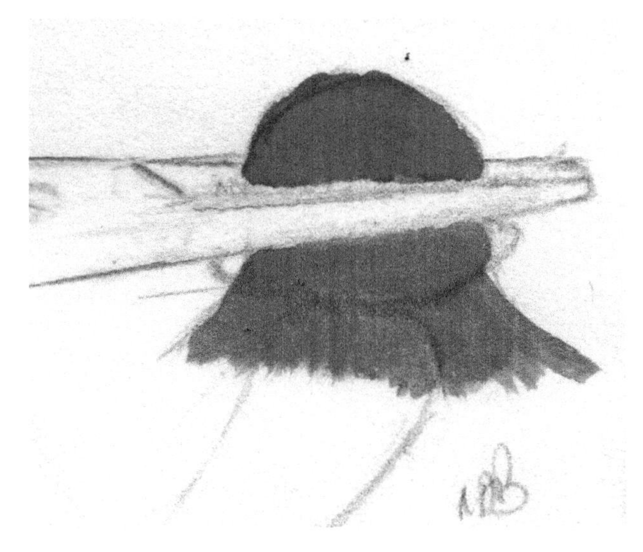

Fig.4.22 (A)

Fig. 4.22 (B)

Clamp radially, ligate (a) and then excise secondary hemorrhoids, (b) before tackling the three major complexes.

Three pairs of large straight artery forceps are employed to set the stage. The "Triangle of Exposure" (Fig 4.23) is displayed before any dissection commences. Straight forceps ensure that the correct orientation at three, seven and eleven o'clock positions is strictly maintained and the dissection is carried out in the same order. The hemostats must not be rotated and throughout they are kept flat, one against the other, in pairs.

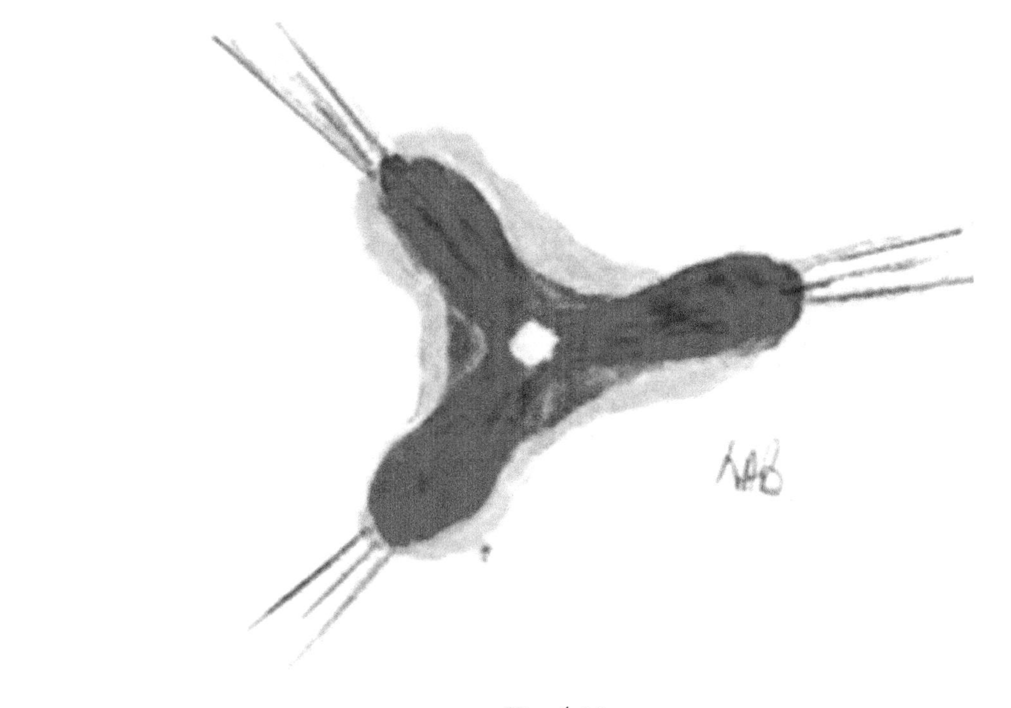

Fig. 4.23

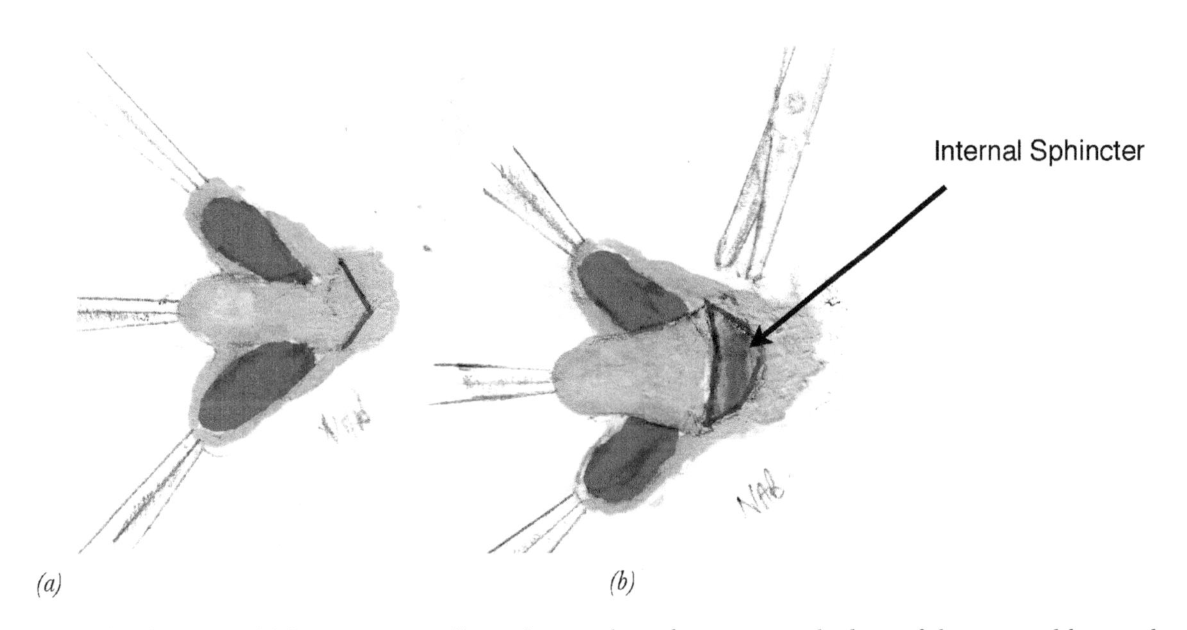

Fig 4.24 The dissection (a) begins externally with a V- shaped incision at the base of the external hemorrhoid mass. (The pedicle forceps are omitted from the diagram for clarity).

This operation is performed entirely extra anally.

The order of excision is constant and always starts with the three o'clock mass. The dissection starts outside the anus, at the external component, and progresses inwards to the internal component. First the external hemorrhoid is clamped and pulled down. This exteriorizes the internal component, which is next clamped at the pedicle. The two clamps, held parallel against one another are then temporarily secured to the drapes with a towel clip.

The two other main hemorrhoid complexes are dealt with in similar fashion. The "Triangle of Exposure" remains clearly displayed and secondary hemorrhoids are also clearly identified.

First the secondary (daughter) hemorrhoids are ligated with 3/0 silk. Don't ligate or excise too vigorously. At every step ensure that you are leaving behind a good amount of ectoderm. The three primary piles are depicted at 3, 7 and eleven o'clock. Two daughter or secondary piles (DP) are so indicated.

You must constantly maintain the orientation of 3, 7 and 11 throughout the procedure. The sequence of dissection follows the order of application of the paired hemostats.

The dissecting instrument is a heavy scissors, such as a Mayo. Do not use a scalpel. It is sometimes helpful to supplement this with gauze dissection- a small gauze sponge is wrapped around your index finger.

Starting with the left lateral pile mass (at 3 o'clock), whilst traction is applied outwards and towards yourself, scissors dissection is started on the lateral aspect. With practice, the dissection is rapid and precise and takes only a few snips. Holding the artery forceps as shown facilitates dissection, which is next directed medially. By separating clamps in the manner shown, (do not hold them in apposition) the submucosal/ subcutaneous space is clearly displayed and maintained throughout the dissection.

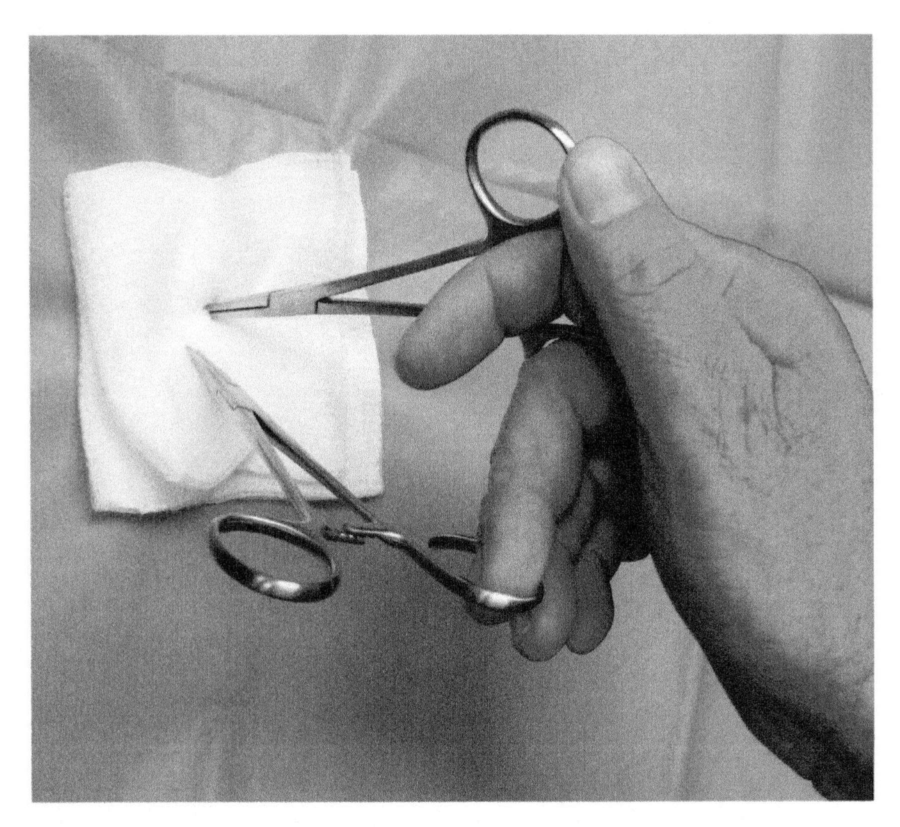

Fig 4.25 During dissection, by holding skin and pedicle clamps splayed out in the manner shown here, you ensure separation and thus constant identification of the submucosal space and thus precise and clean dissection planes.

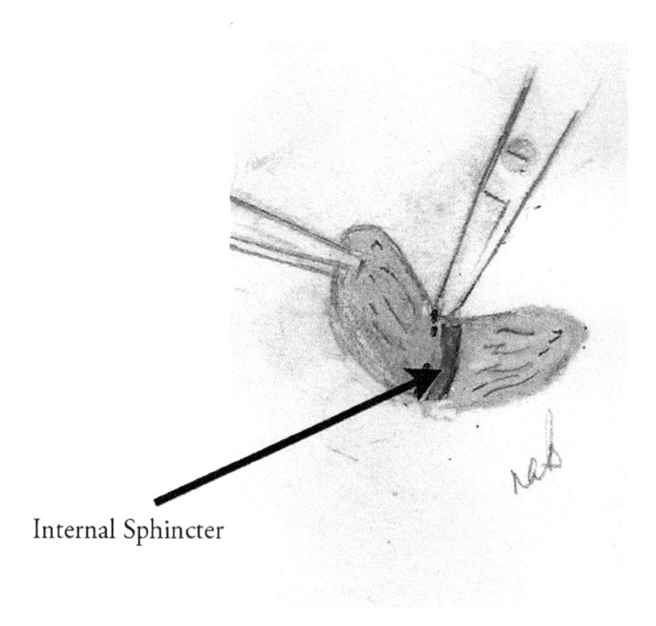

Fig 4.26. The pedicle (mucosal) clamp has been deliberately left out for purposes of clarity. The dissection from outside is continued until the lower margin of the internal sphincter is reached and displayed.

The lower, free border of the internal sphincter is then easily identified together with terminal vertical filaments of the conjoined muscle. Further dissection is then aimed towards the pedicle clamp and stops just inside the internal sphincter. It is not necessary to continue dissecting higher up.

Do not over vigorously excise anoderm; err on the side of conservatism.

A heavy non -absorbable tie, such as a #2 silk or umbilical cord tape, next ligates the mass. Do not transfix. The pedicle forceps is slowly released during the tying, to ensure a good and secure "bite", and then reclamped three to four mms distal to the ligature so that the tie can be double looped around the mass and can be made secure and tight. Be careful to not include any part of the internal sphincter muscle in the ligature. The pedicle forceps is then discarded and the tie is then temporarily clamped to the towels using the second (skin) forceps, which remains orientated at three o'clock, at its original position.

These steps are repeated for each of the other two main hemorrhoids.

You will probably have to change hands on one of the piles to maintain a clear view at all times.

At all times ensure that you are leaving a generous amount of anoderm between each raw wound.

The three ligatures are finally all spread out, maintaining the original 3, 7 and 11 o'clock orientation and pinned to the drapes with one of the discarded clamps (Fig 4.27).

Fig 4.27

Next free the ties, one at a time, in the order of 3, 7 and 11 and excise the ligated masses in the same order (Fig 4.28). Leave the pedicles sufficiently long to ensure non slippage of the ties. The ligatures are then cut to a secure length, immediately after excising each pile (Fig 4.29).

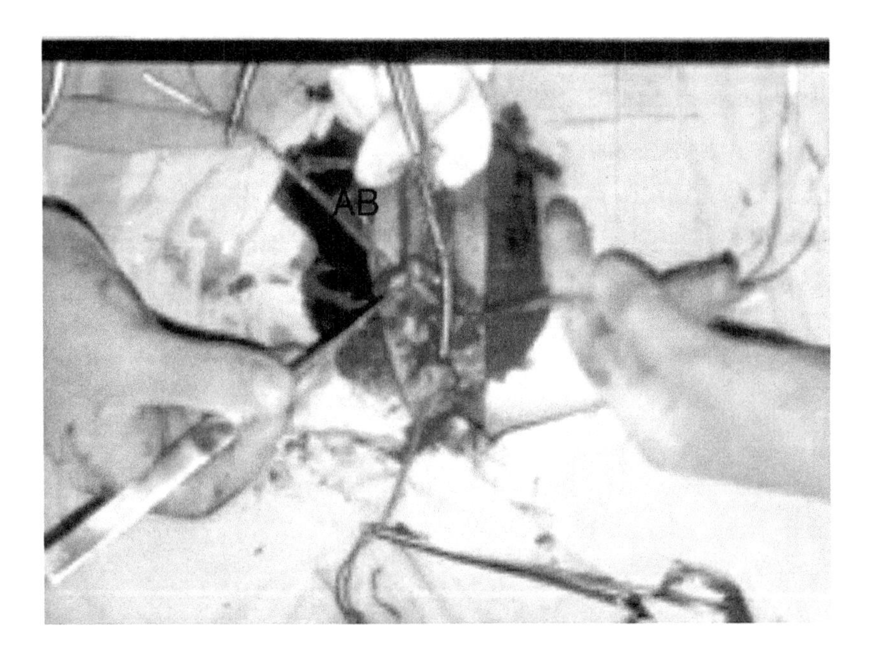

Fig 4.28 Excising the main hemorrhoid masses

Fig 4.29

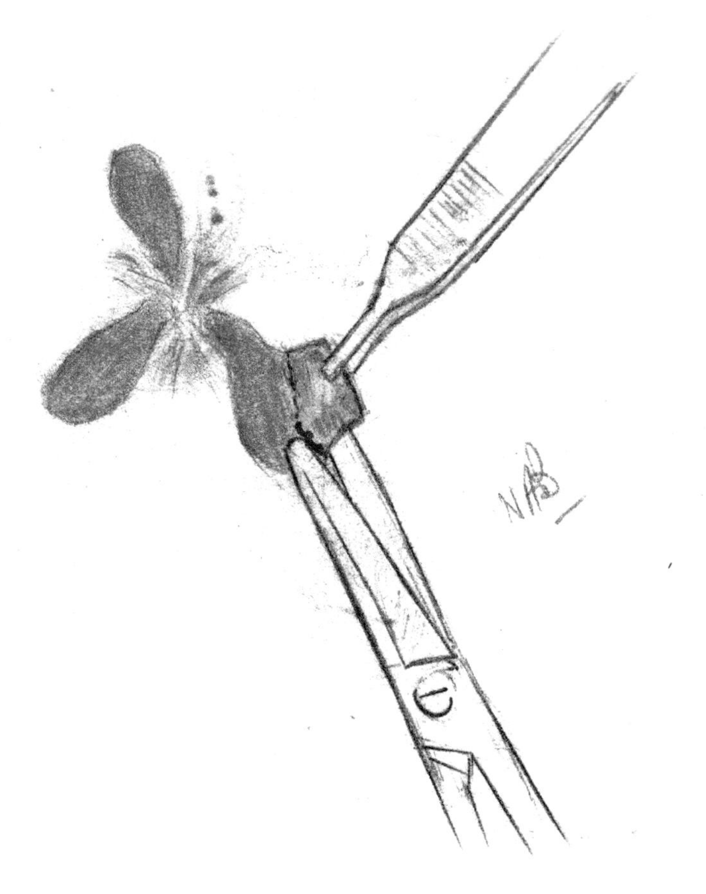

Fig 4.30

Fig 4.31 The final appearance is of a flat anal verge and the raw areas are minimally visible with good intervening skin and anodermal bridges.

All clamps and forceps have now been discarded. The cutaneous wounds are next appropriately trimmed to ensure that there will be no residual troublesome tags.

Trimming residual redundant skin and tags:

On occasions the external tags are so extensive, possibly even circumferential, that you will have to exercise exquisite judgment as to how extensive an excision is safe. Do not perform a totally circumferential perianal excision in such circumstances. Hemostasis is carried out with a Bovie (diathermy coagulation) or fine absorbable ties.

In the very unusual event of symptomatic small residual skin tags, these can be snipped off in the office under local anaesthesia, after several months.

Gently push the ligated hemorrhoid stumps into the anal canal. The skin wounds are left unsutured, to allow drainage and healing by secondary intention.

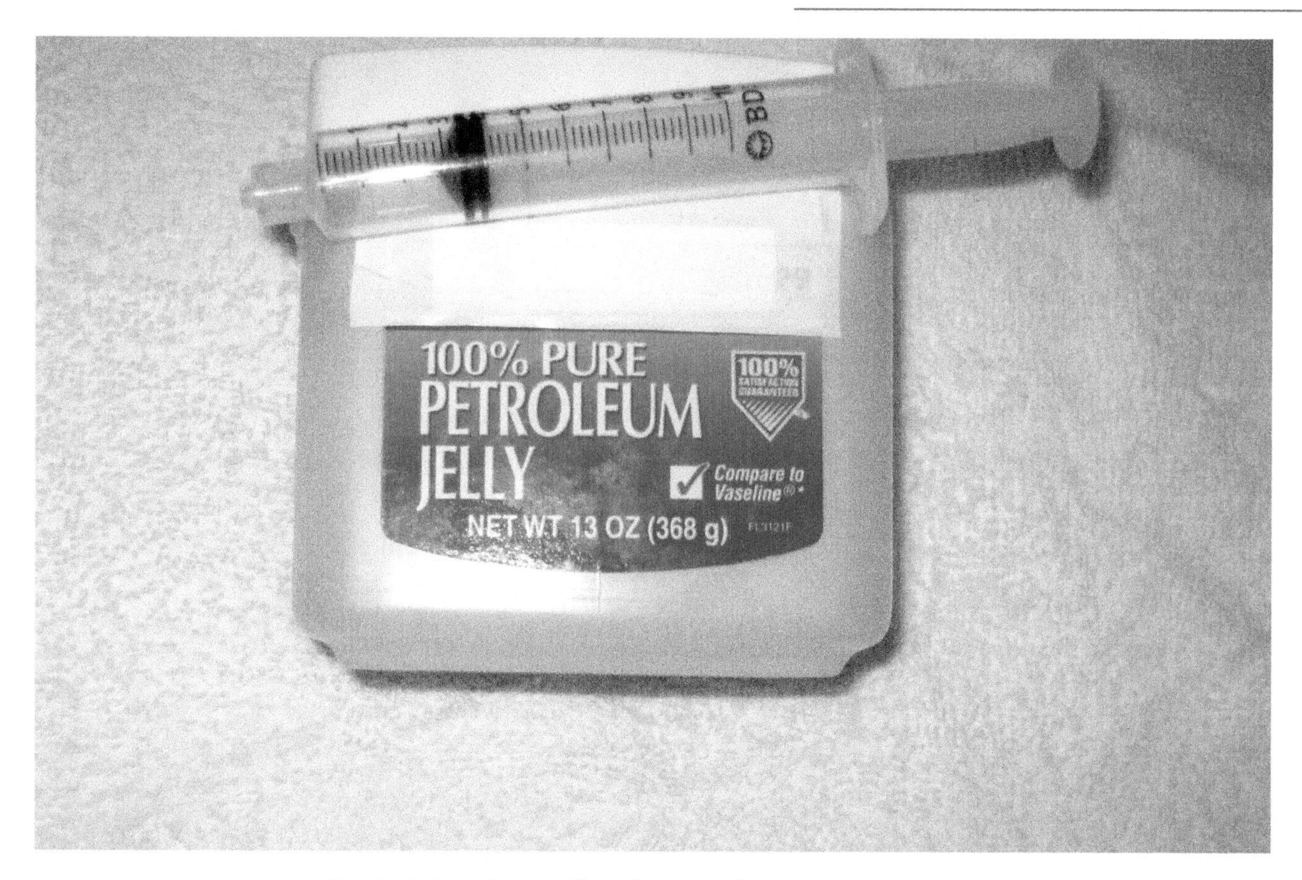

Fig 4.32 Petroleum jelly to be loaded into a 10.0 ccs syringe

Bleeding should never be a significant problem but if you so wish, you may run a loose continuous subcutaneous suture of 4/0 gut for absolute hemostasis. This is rarely necessary but it may have aesthetic appeal.

(It was the author's practice, at this point, to routinely inform the anesthetist that the operation would be over within a few minutes).

Now perform a digital examination. Confirm that there is a good wide lumen and that all pedicles are secure and not bleeding.

A ten cc syringe, (minus a needle!) preloaded with Petroleum jelly is gently introduced into the anal canal and the contents are squirted into the lumen.

Invariably there is a small degree of back leak (because by now the anesthetist is reversing the anesthetic and the patient may be straining). A light dry surface dressing is applied, followed by a T- bandage and the patient is taken out of lithotomy position and the scrotal support tubing is discarded. The patient's calves are vigorously massaged and pummeled and the feet repeatedly alternately dorsi and plantar flexed.

The patient is usually ready for discharge home in six hours, after having urinated.

Follow Up:

There is no place in modern surgery for posthemorrhoidectomy anal dilatations. Once routine, they are painful and unnecessary and they have long been abandoned. Tub baths are taken as desired. Hydrocodone is in most cases adequate for pain control. Excessive medication should be avoided since Codein is constipating . There are no special dressings but some dry pads may be needed for several weeks should there be drainage.

A one time strong laxative is administered on the third postoperative night. You may have your own preference; My routine cocktail laxative (for the first postoperative bowel movement): Mix Cascara, Liquid Paraffin (a.k.a Mineral oil) and Milk of Magnesia each in 1/2 oz quantities. The patient is thereafter kept on a stool softener for three to four weeks, as needed. He/ she is warned about possible bleeding and not to panic. They should expect passage of the silk ligature fragments, although the latter may go unnoticed.

A printed instruction document is given to the patient on discharge. The first follow up visit is in two weeks, when a gentle digital exam is patiently performed. Take your time. It should not be unduly painful if you are patient and gentle. The second and usually final follow up is at six weeks at which time healing should be complete. Rectal exam by then should be painless, absence of stenosis is expected and noted and the patient is instructed to contract the sphincter muscles over the examining finger. This is documented.

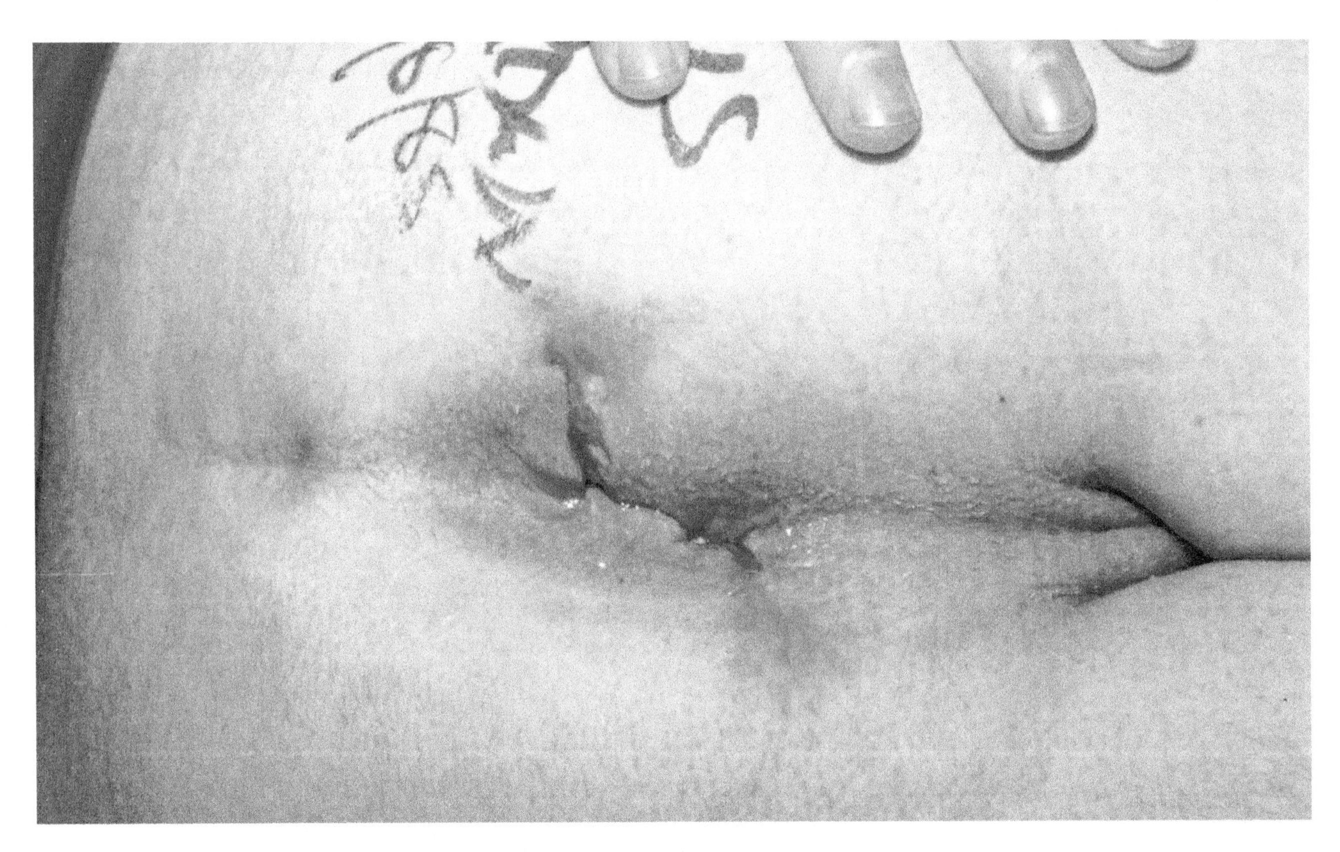

Fig 4.33

2 weeks post operative showing healing with flat perianal area and clean wounds.

Comment

Using this basic technique for some fifty years, the author never encountered a single case of postoperative stenosis nor a single case of incontinence. The sphincter stretch (which was added some forty five years ago) plus minimal excision of anoderm, markedly reduce or may occasionally even completely eliminate the notorious pain associated with hemorrhoidectomy.

Alternative Operations

The Fergusson/Fansler procedure:- This is a socalled closed method of hemorrhoidectomy, performed endo-anally. It is popular in the U S A. Theoretically it has the advantage of absence of distortion of the anal canal during the procedure. The patient is positioned prone (face down) so that the positions of the hemorrhoids are reversed on the clock face; in this position they are located at 1, 5 and 9 o'clock respectively. The operator is standing. Break the table to improve access and then raise it to a comfortable height.

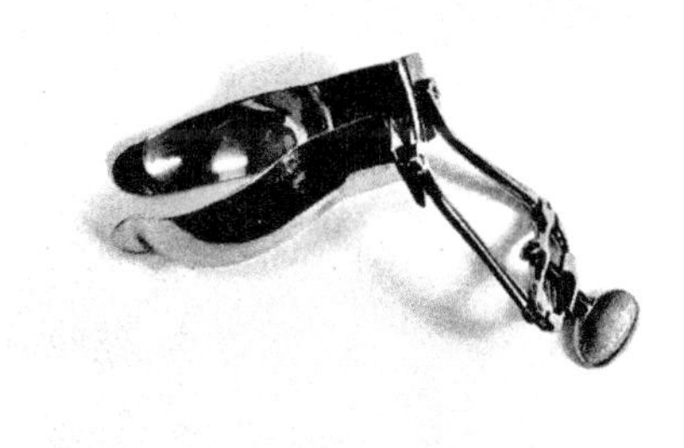

Fig 4.34 Parks Mark 1 retractor

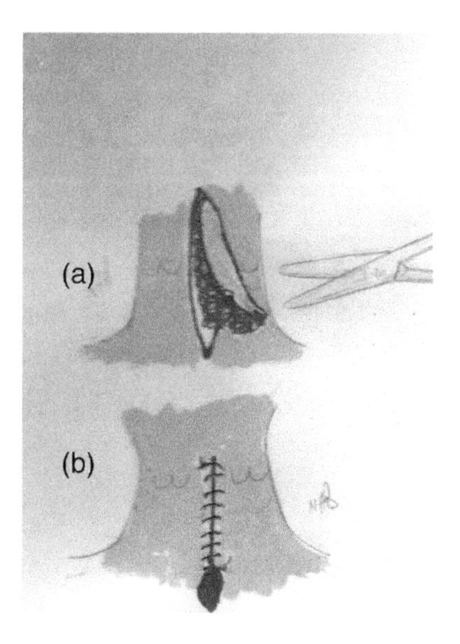

Fig 4.35

An appropriate speculum is inserted. It will have to be re- rotated into appropriate position for each localised hemorrhoid mass. The intra anal speculum stretches and widens the hemorrhoid masses and, in my experience, tends sometimes to blur the lateral margins of the hemorrhoid.

Infiltrate each individual mass with local anaesthetic in epinephrine. Do not distort the hemorrhoid by over filling. You may dissect with a scissors or scalpel. The author found it easier to first define the lateral margins of the hemorrhoid by scalpel dissection and then convert to scissors.

Each of the three hemorrhoid masses is treated identically. The area of the pedicle is first transfixed and underrun with a suture/ ligature of # 1 or #2 chromic gut, which is securely knotted. Next, incise along each lateral border of the hemorrhoid mass from above the dentate to (and including where needed) the perianal skin. The entire composite mass of supradentate mucosa, anoderm, skin and vascular components is easily dissected off the underlying internal sphincter as depicted in 4.35 (a). The wound is then repaired with a running absorbable suture, as in (b).

Stephen Eisenhammer practised an interesting variation which served 6 the double purpose of suturing and simultaneously lifting the mucosa into the anal lumen. Starting at the top of each excision, as each segment of a running suture is placed, it is re -knotted to the starting end. This effectively "lifts" the mucosa upwards and inwards. Alternate bites may include the internal sphincter.

There is minimal bleeding but here and there you may find it necessary, to suture ligate with fine absorbable material or electro coagulate.

The mucosa is approximated with a running suture of 00 catgut (b), starting at the top and leaving a small open area distally for drainage.

Now rotate the retractor to expose the next mass . This is repeated until all three have been removed.

Ligate daughter piles as in the open operation.

After digital examination and establishment of hemostasis, fill anal canal with petroleum jelly (Vaseline) from a syringe.

Comments

This procedure is quite time consuming. The author personally found it "fiddly". The stretch effect of the speculum blades often obscures the precise boundaries of the hemorrhoid.

In my opinion this approach can easily lead to excessive removal of skin and anoderm. The uppermost point of dissection goes much higher than in the Ligature and excision method.

I am not convinced that there is less pain.

You, the operator, have to stand bent over the operating table and it may be hard on your back. It is however possible to perform this procedure with the patient in lithotomy position and the surgeon seated.

In a study by Golligher's group more than forty years ago, the sutures 7 uniformly broke down within ten days, leaving a raw wound to granulate. Healing may take months.

Whiteheads Procedure

First described more than one hundred years ago, a circumferential ring of anal mucosa including submucosal tissues and blood vessels is excised and the resulting defect is repaired by suturing proximal mucosa to the distal anoderm with a series of horizontally placed sutures. Walter Whitehead stressed in his original description, that dissection must be confined to the supradentate epithelium and that the dentate line should not be transgressed. This proved difficult in practice and the operation was technically quite demanding.. I suspect also that many suture lines disrupted, leaving large circumferential intra-anal raw areas to to slowly heal , resulting in severe scarring and an unacceptable incidence of severe post procedure anal stenosis. Furthermore it failed to address the problem of coexisting external haemorrhoids or tags.

Although popular at first it was later abandoned for the above reasons. Some fifteen years ago there were attempts at revival. Some authors were quite enthusiastic but it did not gain widespread support and it has once again faded.

Submucosal Hemorrhoidectomy

The late Allan Parks refined the Fergusson operation by incising only through the overlying anoderm and mucosa, then burrowing laterally to raise mucosal flaps for a short distance and removing only the vascular submucosal tissues. This was followed by suture repair of the anoderm.

In my experience this operation was "fiddly" and time -consuming. It was included in Golligher's study which demonstrated the inevitable breakdown of the suture repair.

It did not enjoy widespread use and it appears that it is now rarely if ever performed.

Stapled Hemorrhoidectomy

This innovative procedure was introduced in the late nineties. The author has no personal experience of same but it is clear that the problem of the external components is not addressed. Furthermore other problems are reported.

Stapled hemorrhoidectomy is clearly a sophisticated modern day version of the Whitehead procedure and would likely interest those surgeons who subscribe to that approach. It has the appeal of apparent ease of performance and precision and is alleged by its advocates to be less painful than traditional excisional hemorrhoidectomy.

It involves the circumferential excision of supra dentate mucosa but cannot address the issue of significant, fibrosed external hemorrhoids or tags. I would venture that it may have a place in treating moderate circumferential ano-rectal mucosal prolapse.

I have difficulty in accepting studies on pain intensity since there is no method whereby it can be scientifically compared objectively between individuals, whose pain thresholds may differ vastly.

Using simple guidelines such as frequency, dosage and nature of analgesia and also based on my personal experience as a patient, I am convinced that the intensity of post hemorrhoidectomy pain is grossly exaggerated (with the single exception of the first bowel movement), especially when combined with the technique outlined and in particular with anal stretching. After the introduction of "managed care" in the USA, for more than twenty years in the author's practice, more than one thousand haemorrhoidectomies were performed on outpatients except for two. Of these, one was hospitalised overnight because of post operative urinary difficulties and one for pain medication. Mild analgesia was adequate in all other cases.

CHAPTER 5

POST HEMORRHOIDECTOMY ANAL STENOSIS

This most disabling condition is the end result of a badly performed hemorrhoidectomy where there has been excessive removal of anoderm.

Operation

In lithotomy position. Under General Anesthesia , stretch the anus. There will likely be some bleeding from torn scar tissue.

Insert a double bladed speculum. Incise the scarred lining (proctotomy) at seven or eight o'clock . As the scar is divided , the lumen yields and so allows the speculum blades to be gradually separated further. The incision should be deepened until the underlying smooth internal sphincter is displayed. An additional limited lower internal sphincterotomy may be needed. The anal lumen should now comfortably admit two fingers.

With ink, mark the outline of the proposed peri -anal fat- cutaneous advancement flap, the size to match the defect that you have just created. Since survival of the flap will depend on a tenuous blood supply via the underlying fat layer, help ensure as you mobilize the flap that its pedicle remains broader than the skin island.

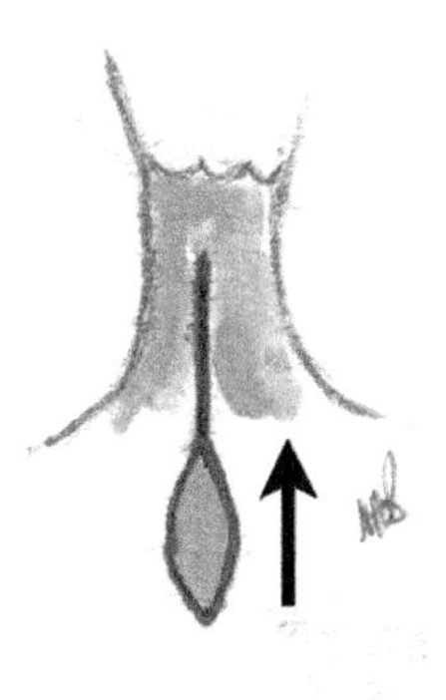

Fig 5.1 Release of scar and creation of flap. The arrow designates planned later advancement of flap.

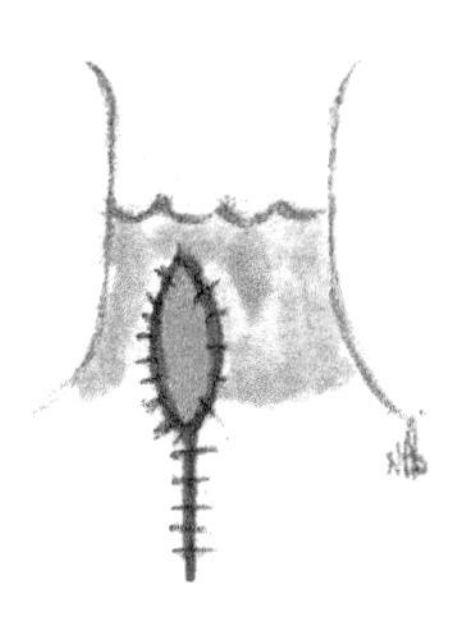

Fig 5.2 Flap has been transposed and sutured in place. The Donor site has been sutured.

Figure 5.3 The base of the flap is fashioned to be much wider than the skin strip to help ensure a good blood supply.

It should easily transpose into its new position, without tension.

Secure it with fine (4/0 or finer) interrupted Vicryl sutures all round. Do not tie the knots over tightly.

Suture the donor site with fine silk interrupted sutures, for subsequent removal. If need be , leave the inferior extremity of the donor site unsutured.

Inject Vaseline into the canal and apply a dry dressing.

Although one procedure will probably suffice, if needed depending on the severity of the stenosis, the procedure can be repeated three months later, this time at the four or five o'clock position.

There is no postoperative disability and the patient can return to work after a day or two.

CHAPTER 6

ANAL FISSURE

This is a radial crack in the anoderm, situated at the anal verge, usually in the midline dorsally and directed into the Anal Canal. It may be idiopathic or secondary to underlying disease. It is quite superficial and is easily displayed, when gently separating the margins of the anal canal by spreading the buttocks. Do not force your index finger into the anus to demonstrate a fissure. This is unnecessary and it can be intolerably painful.

There may be a small overlying external "hood', which is sometimes called a "sentinel pile". Where it has followed passage of constipated stool or severe diarrhea the etiologic provoking factor is clear. Often however there is no clear cut etiology. It is important to be aware that a fissure may be a manifestation of disease higher up, such as Crohn's Disease, Ulcerative Colitis or Chronic Diarrhoea from any cause. Multiple and non-midline fissures suggest underlying primary bowel pathology.

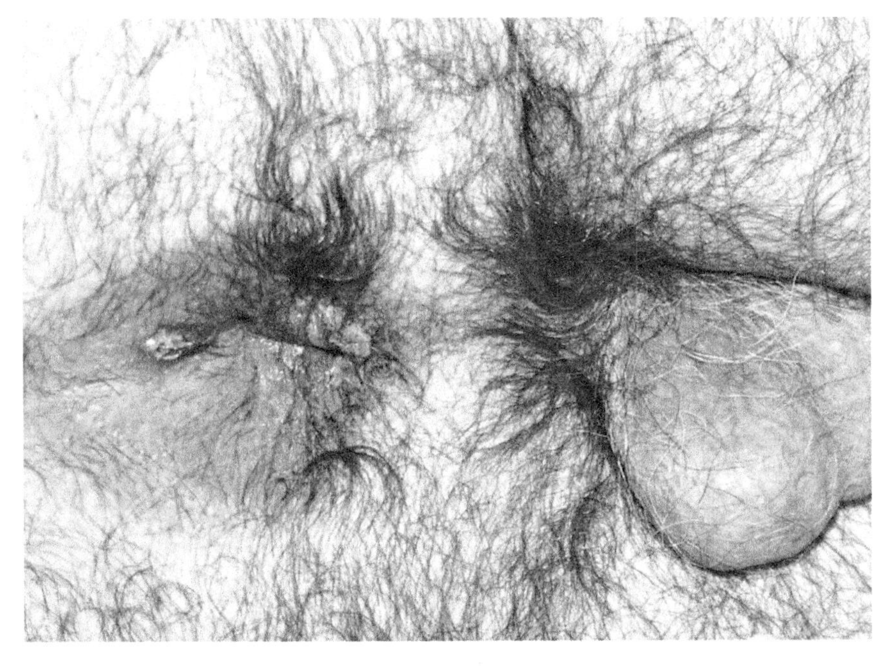

Fig 6.1 The fissure is surmounted by a dorsal hood.

Anal fissure can be extremely painful because of intense underlying sphincter spasm, especially during defaecation. The stool tends to be "pinched off" and its passage may also be accompanied by fresh bleeding. This may be seen only on the toilet paper, on the side of a formed stool or in the toilet bowl as well.

Fissures are either acute or chronic. The latter will likely require surgery; it is indurated and each act of defecation tends to split the tissues and prevent healing.

Sphincter spasm maintains the fissure and a vicious cycle is thus established. The idiopathic fissure is usually single and located in the midline dorsally in some 90% of cases, 10% in the midline anteriorly and rarely laterally.

I suggest the possibility that the patient may have passed flatus perhaps during sleep, against an uncoordinated, unrelaxed sphincter.

Treatment

Acute fissure - will respond to local hygiene and Local Anesthetic ointment. The patient should be on stool softeners.

Chronic Fissure - General anaesthesia is preferred. The time honored effective cure is by partial internal sphincterotomy. If the patient has not already undergone fibreoptic endoscopy, sigmoidoscopy, rigid or flexible, is then also performed.

The surgical options are:

- Anal Sphincter Stretch
- Dorsal midline open Sphincterotomy:
- Lateral subcutaneous Sphincterotomy

Anal Sphincter Stretch

Overstretching (Lord Procedure) may result in incontinence of varying duration. Understretching may fail, although short duration relief may be achieved. Either way, recurrence is practically certain.

Botox Injection

This was first hailed as a minimally invasive alternative treatment. After several years it is now clear that the benefit is temporary and recurrence is the rule and this is no longer considered to be a serious option.

Lateral Subcutaneous Internal Sphincterotomy

General anesthesia is preferred, with the patient in lithotomy position

First perform careful bidigital examination, rotating the fingers from one side to the other and back. If you apply soapy solution to your gloved fingers this imparts great sensitivity and you may detect minimal, otherwise elusive pathology. The lower margin of the internal sphincter is made tense by finger pressure. Infiltrate 2 to 3 ccs of 1% local anesthetic in 1:100000 epinephrine at 3 o'clock over the palpable depression. Then massage to disperse.

Apply gentle bidigital massage to disperse fluid. Maintained pressure from your left index finger helps identify the muscle. If you have any doubt, a small bivalve speculum modestly opened will facilitate identification and accurate division of the sphincter.

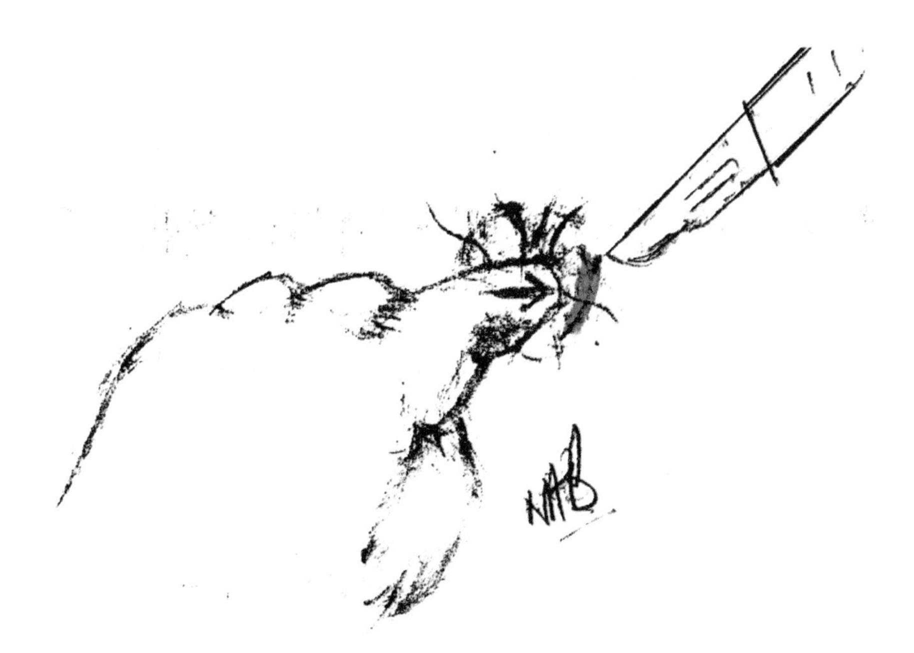

Fig 6.2 Using a #15 blade, make a 0.75 cm incision parallel to the anal verge (Fg 6.2), over the distal end of the internal sphincter (palpable depression). Gently splay open with a fine scissors or hemostat, to expose the white colored, distal end of the internal sphincter.

Splaying is continued first submucosally and then on the lateral side of the muscle, to the planned depth of the sphincterotomy. This should equal and not exceed the length of the fissure. Throughout, excercise great care not to buttonhole the anoderm. Splaying the blades of a fine scissors will display the lower margin of the sphincter, which is subcutaneous and is readily recognized from its position and whitish color. Free the lateral and medial sides of the muscle to a depth of approximately 0.5 cm. Exercise care not to buttonhole the anoderm.

There may be some venules coursing on the sphincter edge (fig 6.3).; electrocoagulate them on a low setting. Protect the neighboring skin by gently retracting it out of the way with a Gillies skin hook or Senn's catspaw retractor.

Stabilise the muscle by holding with a Judd Allis (non crushing) clamp. With a scissors, divide the lower part of the sphincter under clear vision, for the desired distance (fig 6.4). Be unhurried and precise. Temporary pressure should control any bleeding; this will soon cease. Finally, inject Vaseline into the canal and apply a dry dressing. With scissors, divide the lower part of the sphincter to the desired depth (to equal the length of the fissure, fig 6.9). Palpate to ensure complete division. Excise the sentinel tag and submit for biopsy. The incision is not sutured; it will heal invisibly in a week or two.

A small snip of the lateral wall of the fissure plus the sentinel tag, if present, are submitted for biopsy.

This completes the operation, which probably took less than 5 minutes. Perform a postprocedure bidigital rectal examination. Inject Vaseline into the canal and apply a light dressing.

There is no special after treatment.
No restriction of activities.

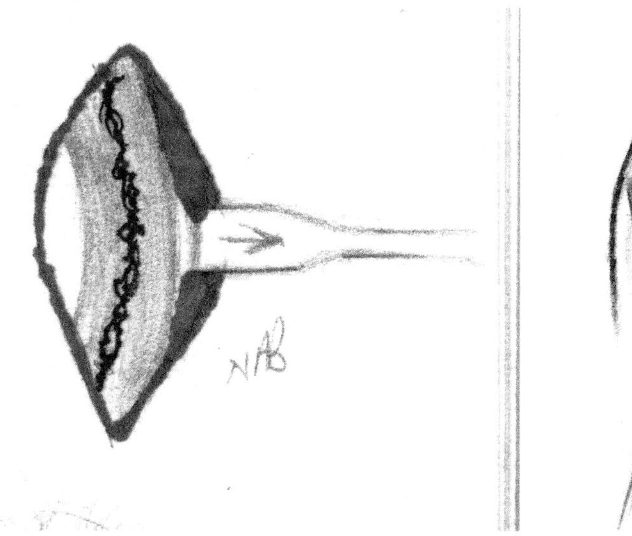

Fig 6.3

Fig 6.4

Dorsal Open Internal Sphincterotomy

This operation fell into disrepute because of the resultant "keyhole" effect and mucoid leak. The problem is obviated by the modified technique described below and there remains a place for this procedure, the original method of Eisenhammer, where there is longstanding fibrotic contracture with significant stenosis or a sizable skin tag (sentinel pile) which may or not be associated with an underlying abscess.

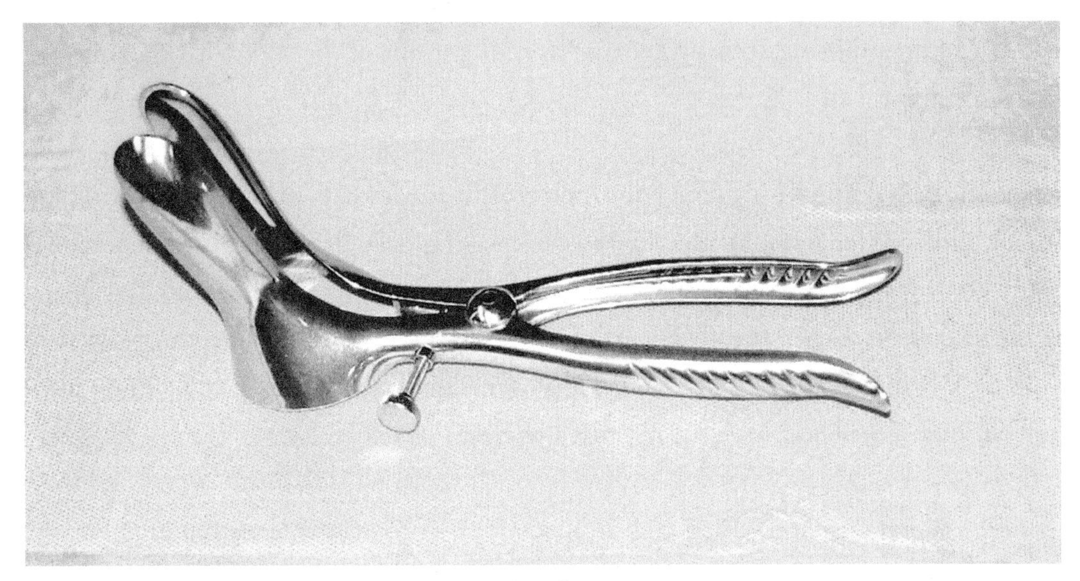

Fig. 6.5

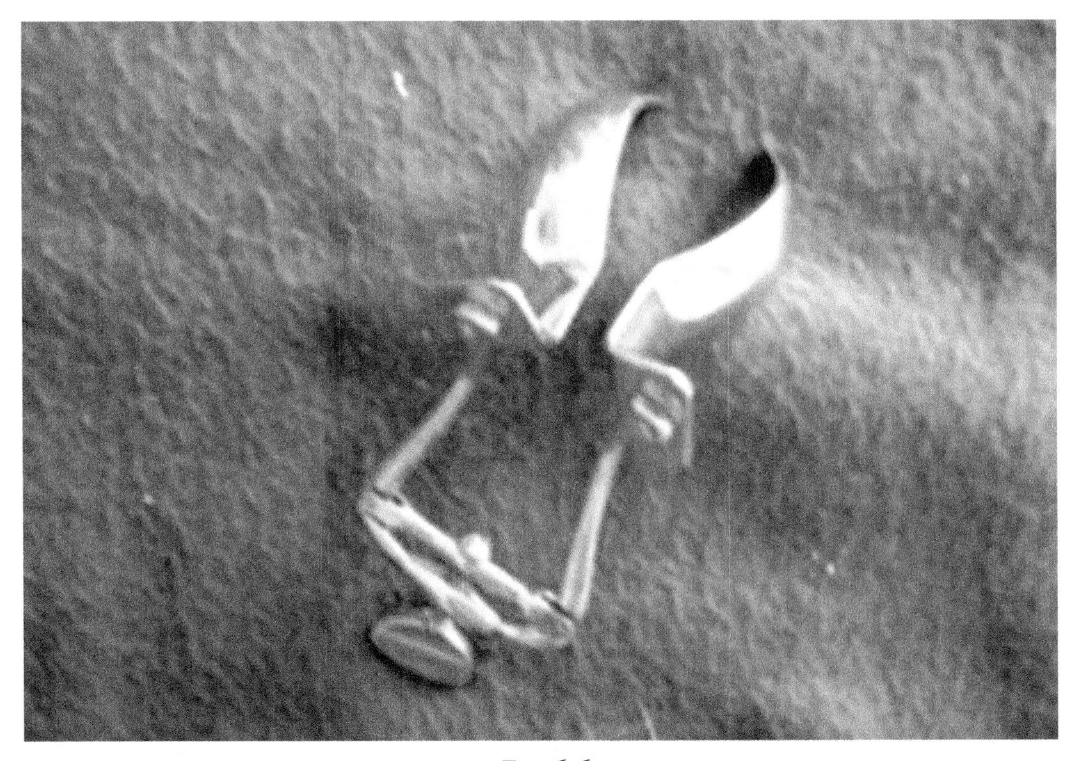

Fig. 6.6

Procedure

After a gentle stretch, insert a double bladed speculum (Fig 6.5 or Fig 6.6). Gentle separation of the blades will open up the fissure and expose its floor, accompanied by some bleeding. This is easily controlled with local pressure. Inject one or two ccs of epinephrine solution into the floor. The horizontally disposed white fibres of the internal sphincter muscle are clearly seen, as they are put on stretch by the speculum.

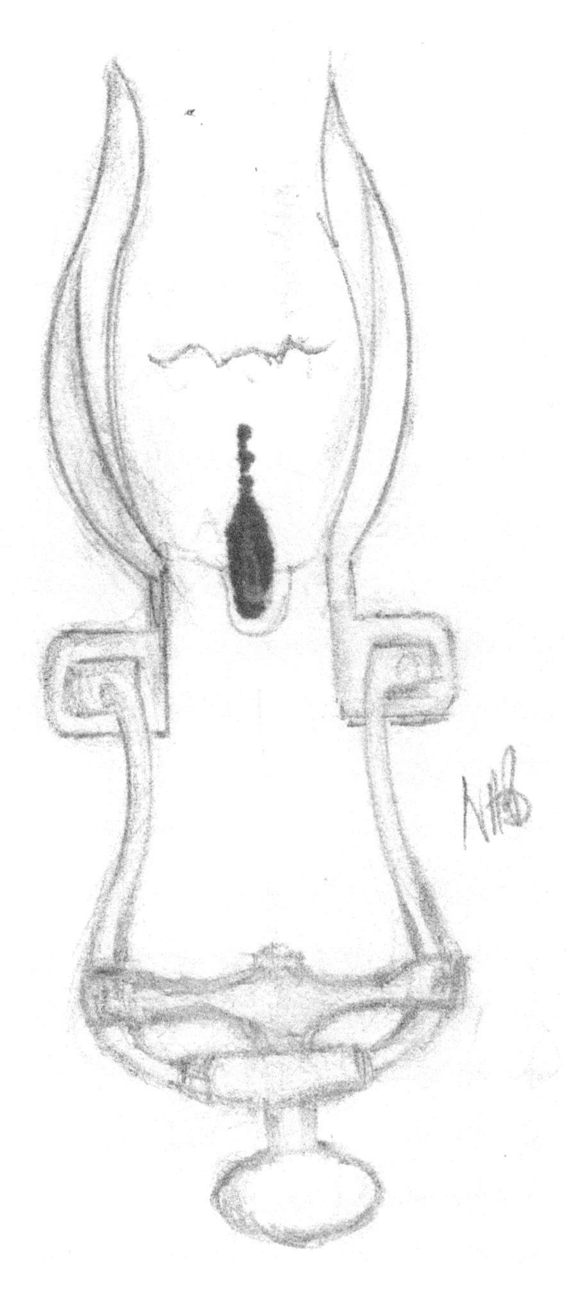

Fig 6.7 Opening the blades of the retractor exposes the fissure in the dorsal midline. The horizontally disposed underlying internal sphincter fibres are clearly displayed. Periodic readjustment enables precise planning of the length and depth of the sphincterotomy;

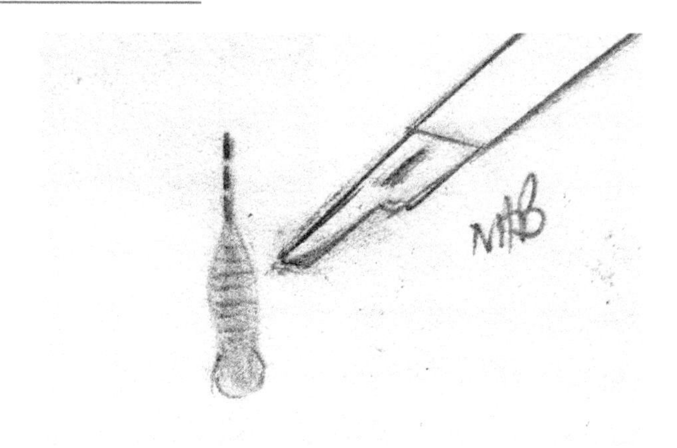

Fig 6.8 Underlying internal sphincter clearly displayed. Showing proximal extension of incision.

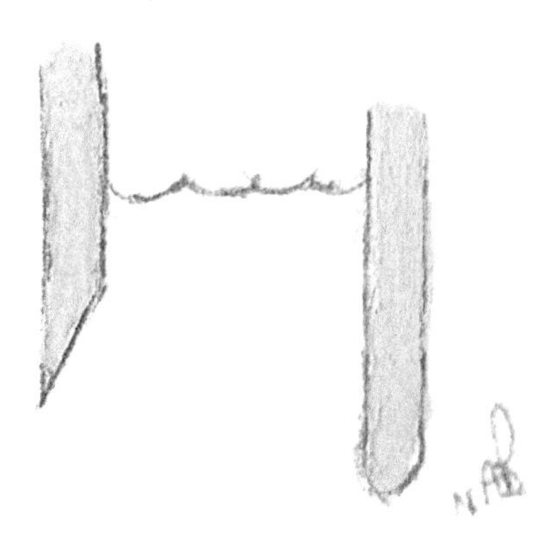

Fig 6.9

Lateral view. With a scalpel, bevel the sphincterotomy by a sloping partial thickness extension proximally to create a smooth transition. Periodic readjustment of the retractor blades makes this very precise and the undesirable keyhole deformity is completely avoided.

A small snippet together with the sentinel tag, is sent for biopsy.

References:

Regarding sphincterotomy for Anal Fissure Norman A Blumberg Diseases Colon and Rectum Vol 41 (8): 1071-2 (1998)

CHAPTER 7

THROMBOSED EXTERNAL HEMORRHOID OR ACUTE PERIANAL HEMATOMA

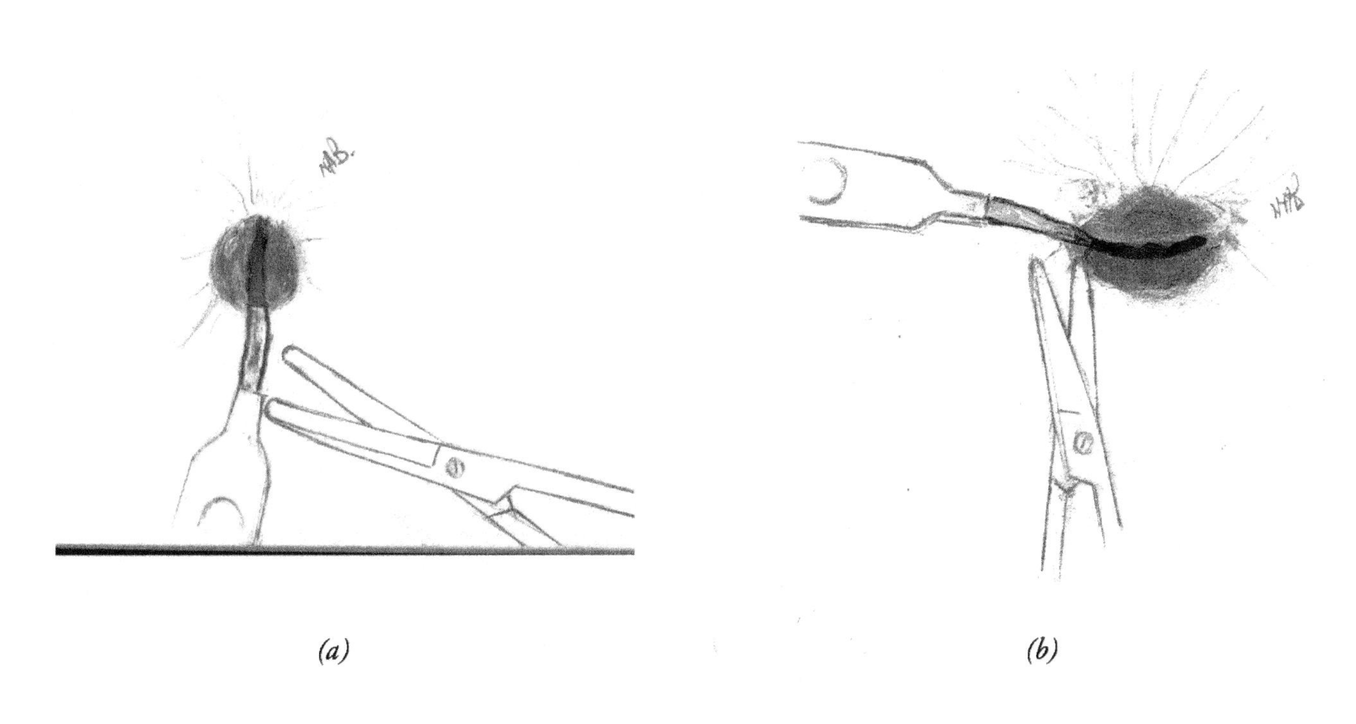

(a) (b)

Fig 7.1

(a) *Perianal hematoma (a.k.a thombosed external pile). Radial incision followed by excision of skin strip.*

(b) *Whether the incision is radial or parallel will be determined by the direction of the swelling. When radial avoid continuing into the anal canal.*

This common, extremely painful acute condition is easily cured with simple surgery, performed in the office, under local anesthesia. The lesion may be an intravascular clot or an extravascular hematoma. Straining at stool is the usual incriminated etiology, although this history is often absent.

For surgery the lateral decubitus position (whether left or right depends on the site of the lesion) is employed; the lesion should be lowermost. Your nurse/ assistant holds the upper buttock out of the way. In the absence of assistance, the patient can hold his or her own buttock. This is effective.

Infiltrate the swelling with local anesthetic . Next grasp the anaesthetised skin at the dome of the swelling with a toothed instrument or hemostat. Using a fine scissors excise an overlying strip or ellipse of the skin. Whether the ellipse is radially directed or parallel to the anal margin is determined by the direction taken by the swelling.

Evacuate a free lying clot by splaying out the scissors blades. If the clot is contained, excise it together with the related local venous channels.

Bleeding will be easily controlled with a few minutes of local pressure. No sutures are necessary.

If the clot overlies a lateral rupture in the venous sidewall (fig 7.2 a), the affected venous segment will require excision (fig.7.2 b). This will allow the divided free ends of the vessel to then contract, retract and seal off. No sutures or ties are necessary. In this situation , if you do not excise the venules, bleeding may persist. With frank rupture, excision of venules will not be necessary.

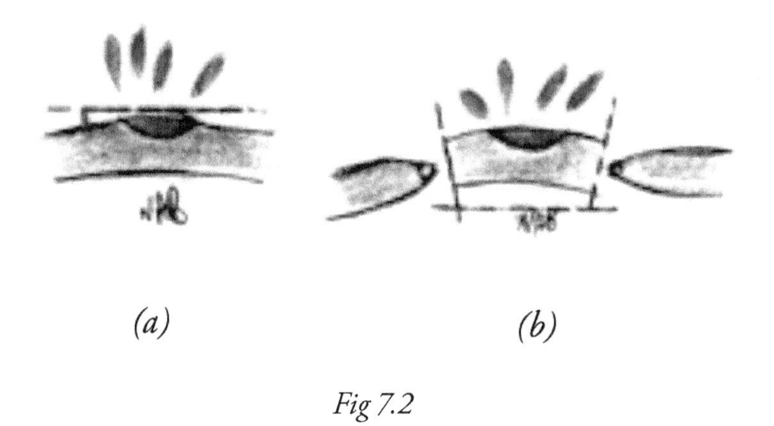

(a) *(b)*

Fig 7.2

Apply a dry dressing.

Relief of pain is immediate. There is no further treatment apart from local hygiene and baths. The wound heals in a few days.

Warn the patient about possible slight bleeding from the site until fully healed.

CHAPTER 8

ABSCESS

Although local abscess may be secondary to underlying anorectal or colonic disease, the majority of abscesses have no demonstrable underlying cause.

It is quite surprising that perianal and rectal abscess are not seen much more frequently. In third world populations however this is not the case. The reason is obscure: personal hygiene has no role in the etiology. Perhaps an important factor may be failure to present early for treatment of minor conditions.

Abscess may occur in any of the spaces represented in Chapter 2, Figs 2.4 and 2.7. The superficial abscesses e.g perianal, are usually easily recognized. The cardinal symptom is intense pain in the area, together with systemic features of infection, including fever, malaise, toxemia etc. The area overlying will be tender, red and manifest the features of cellulitis. Patients may present before there is clear physical evidence of localised pus. Deep seated abscesses such as in the deep postanal space and pelvirectal (supralevator) areas may elude early diagnosis and require a strong index of suspicion and pelvic examination.

Careful bidigital examination will readily identify the abscess site. The site of maximum tenderness is a marker for the site of a surgical incision for drainage.

Note : All abscesses are drained perineally with two exceptions; these are submucosal abscess, which is incised and drained intraluminally, and supralevator abscess. The latter is drained transabdominally and the underlying pathology dealt with as needed.

Some caveats:

Post anal abscess may track across the midline to the opposite side, and present as a horseshoe lesion. This may dictate the need for a second (counter) incision, possibly on the opposite side, for adequate drainage.

Note: Do not procrastinate, while waiting for the abscess to point, before instituting surgical drainage. If you do, a simple abscess may convert to a fistula while you are watching and waiting.

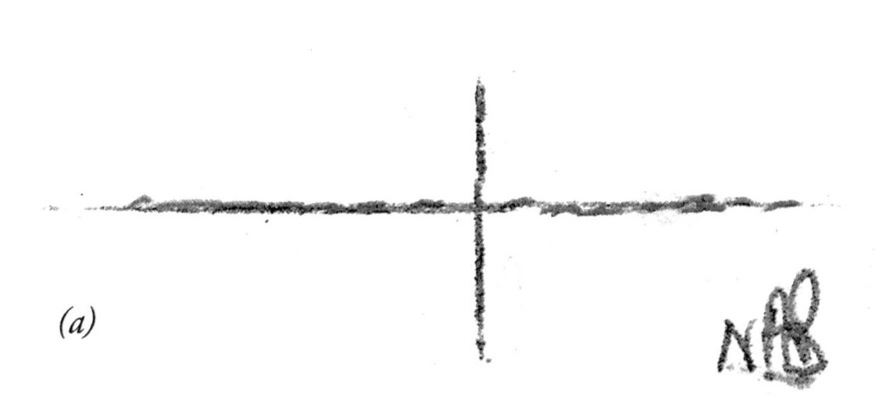

Fig 8.1 Cruciate incision.

Fig 8.2 Diamond shaped defect (b) after excising the corners.

Apply a well padded dry dressing

Choice of Incision:

Patient in lithotomy position, under general anesthesia. Perform bidigital examination. Rotate the fingers, well lubricated with soapy solution, back and forth. Make a one to two inch long incision over the center of the swelling. (Fig 8.1) In the case of Ischio /Rectal abscess, a transverse incision will help avoid injury to the Inferior Haemorrhoidal nerve. Anticipate that pus under tension may squirt out. Therefore exercise care to not receive the jet of infected fluid in your face, by shielding with the dish and the judicious use of the sucker poised and kept at the ready, close by. More than sufficient pus can thus be collected to submit for culture.

Next, with a finger, gently break down any loculi within the cavity and follow with curettage of the cavity with a serrated uterine curette. Submit scrapings for culture and histopathology. Whilst curetting, exercise the utmost care to avoid perforation of the bowel as well as sphincter damage. Be extremely careful when the curette is close to the bowel wall. A finger kept simultaneously inside the canal imparts sensitive appreciation and will assist in safeguarding the bowel and sphincters. The incision can now be extended if indicated. Because the abscess may be associated with an anal fistula which will likely require later elective treatment, try to picture the course of a possible track according to Goodsall's rule. You may look but do not search or probe for a fistula at this time. Leave that possibility for a later date. A 1/2 inch incision is next made at right angles across the center of the primary incision and the corners excised (Figs 8.1 and 8.2) leaving a small diamond shaped defect. This will help ensure against premature closure and allow for necessary prolonged drainage. A corner of a gauze sponge is lightly tucked into the cavity This is then covered with a multi-layered, absorbent gauze dressing, secured with paper tape, loosely applied to neighboring skin. Tip: Never apply paper tape anywhere under stretch. It is merciless in tearing the skin.

There is no special after treatment apart from local hygiene and baths.

Follow up at fortnightly intervals.

The wound should close within a few weeks, from within out. If drainage persists after several weeks, this will suggest a probable underlying fistula and gentle probing is then permissible.

CHAPTER 9

FISTULA IN ANO

By definition a fistula is a track connecting two epithelial surfaces, in this case between anal epithelium and perianal skin.

Chronic anal fistula was first recognised and so named by Stephen Eisenhammer, as an "Anal Fistulous Abscess" . The concept is that infection starts in an anal intermuscular 8 gland and it then burrows in one of a variety of directions. Internally it communicates with the anal lumen at the dentate line, in a crypt, behind a cusp. This opening is therefore well below the ano rectal ring. It follows that for a fistula to heal, the area between the internal and external sphincters will have to be cleared out of smoldering infected anal gland tissue. Any internal opening higher than the dentate line is almost certainly the result of previous surgical trauma.

The position of the external opening varies. There may be a single or multiple such openings, especially after multiple repeated episodes of abscess formation. They will all eventually intercommunicate via a single underlying track. The course and possible varying levels of the track are of tremendous importance to the surgeon, because irrespective of the site of the internal opening, surgery on the main part of the track may endanger the deep part of the sphincters, including the ano rectal ring. You must therefore think two dimensionally about the course of a fistula; both horizontally and vertically.

Where there is one external opening, Goodsall's rule is a useful guide to the direction of the track. The rule states that an imaginary line is drawn transversely across the anal orifice, thereby dividing same into anterior and posterior halves, and if the external opening lies anterior to this, the track will pass radially and directly between the opening and the anal lumen. If the external opening is posterior to the line, the track will curve posteriorly and communicate in the dorsal midline.

This is however not invariable. The corollary is: If the external opening lies within a one inch wide imaginary ring surrounding the anal orifice (colored yellow in the illustration) the track may run radially whether anterior or posterior to the transverse line (Fig 9.1 A and B) . Similarly, if the external opening is outside the colored zone, whether anterior or posterior, the track may curve and it may communicate in the midline dorsally (Fig 9.1 C).

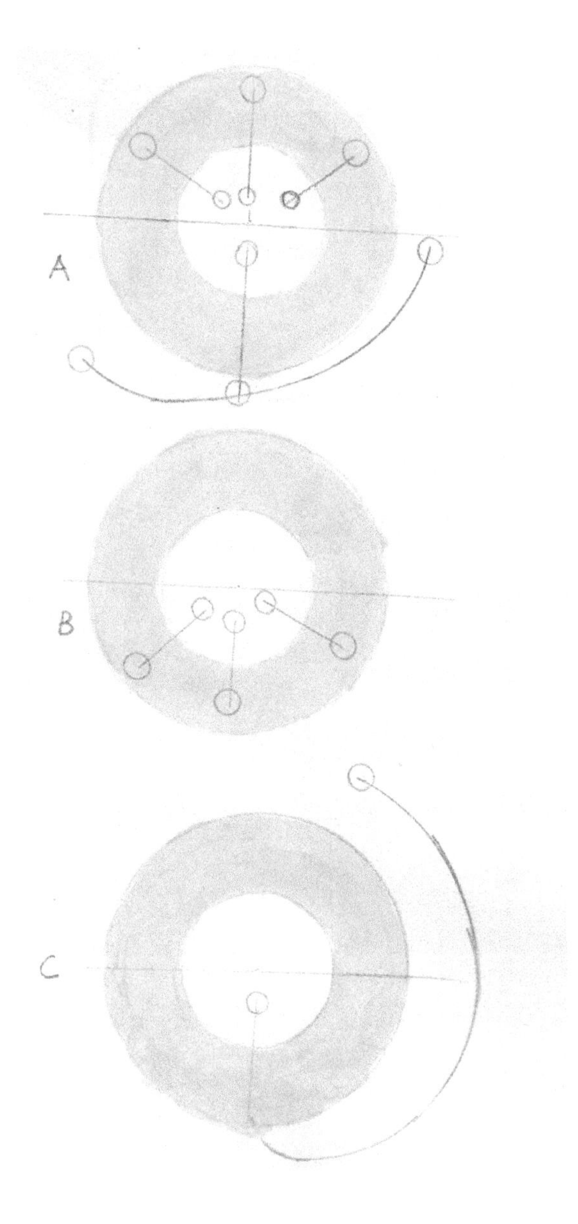

Fig 9.1 Illustrating Goodsall's Rule (A) and Corollary (B and C). The yellow doughnut represents a theoretical one inch wide circumanal zone.

The practical implication is that if the skin opening is within the doughnut, the track will almost certainly be short and direct and division of the subcutaneous part of the external sphincter will not result in disability or incontinence but where there is an opening outside the doughnut, the course of the track will be more complicated and sphincter division may be problematic. In the author's experience, where the external opening is outside the doughnut, the fistula tends to be higher.

The majority of anal fistulae result from a simple abscess in the area or previous local surgery. A small number occur secondary to underlying anorectal disease, e.g. Crohn's and Ulcerative Colitis. Tuberculosis is sometimes seen in the third world. In these cases the local treatment of the fistula may be more demanding and complicated and in addition require appropriate treatment of the underlying, primary disease.

Evaluation:

Perform a thorough bidigital examination. You may have to switch hands to completely evaluate the area through 360 degrees. This allows precise anal, perianal and anal space evaluation. Soapy solution on the gloves imparts exquisite sensitivity during to and fro rotation.

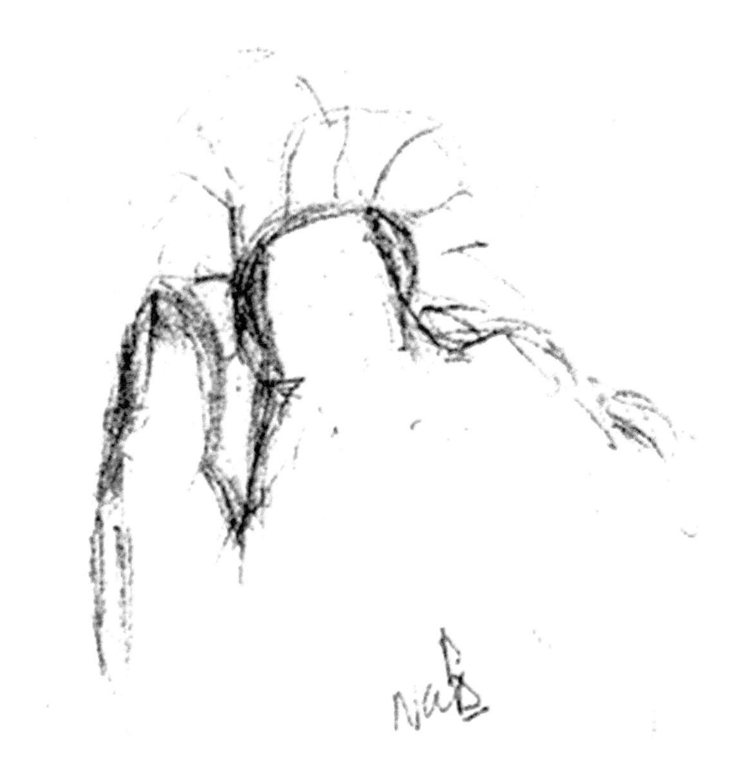

Fig 9.2 Bidigital examination. Thumb at palpable depression (distal limit of intermuscular space).

A working guide for preservation of rectal continence after fistulotomy.

Although the internal opening will expectedly in most cases be at the dentate line, the actual classification of anal fistula relates to the level and course of the track itself in relation to the subdivisions or components of the external sphincter. There are numerous variations and reclassifications of such fistulae. In practice these are a guide to the safety of open fistulotomy with regard to anorectal incontinence.

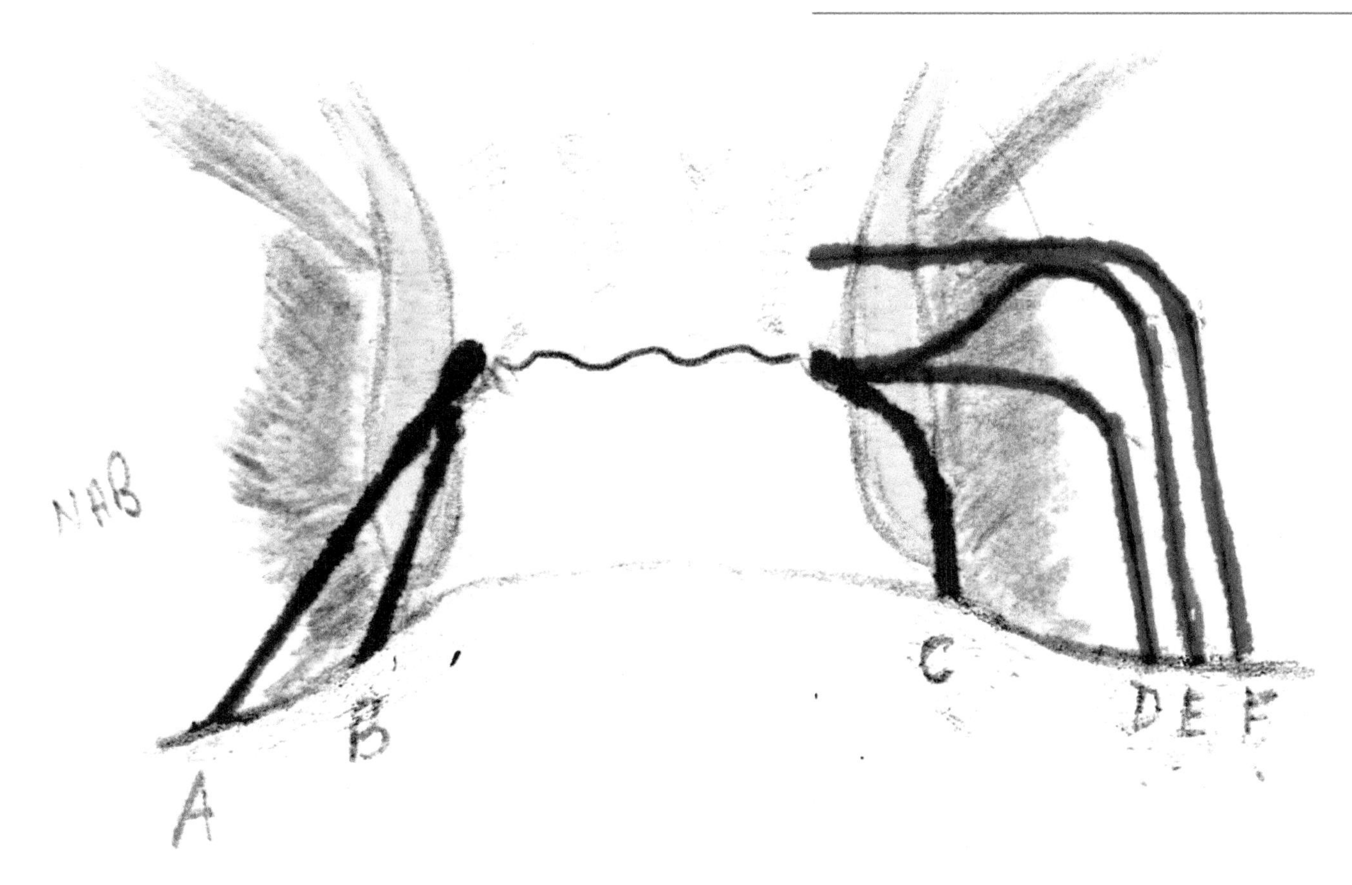

Fig 9.3 In situations A, B and C (highlighted in green) which transgress the superficial external sphincter or the lower half of the intermediate external sphincter, fistulotomy is safe. In situations D, E and F (highlighted in red) which transgress the upper half of the internal/ intermediate external sphincter complex , fistulotomy risks incontinence.

As a rule of thumb, the closer the surgery gets to the ano -rectal ring, the higher the risk. Incisions made for subcutaneous and low transmuscular fistula do not threaten continence. High transmuscular fistula the author believes to be beyond the province of the occasional general surgeon and you will have to critically evaluate the situation, including consideration of placement of a Seton, or redirect the patient to a spcialised high volume colorectal surgeon.

In summary, division of the intermediate part of the external sphincter for more than half its height threatens continence. If the track crosses the intermediate part of the external sphincter anywhere below the level of the dentate line (approximately 5/8 inches from the anal verge) , fistulotomy is safe. If however the track passes at a higher level or pursues a devious course leading to division higher than the dentate line, incontinence may or will likely ensue. " The operations

Fistulotomy. The time honored Procedure

You have started out with a concept based on the location of the external opening.

Bidigital examination with well lubricated fingers yields important information and guides your next step. You must assess the caliber of the track and choose an appropriate size and shape of probe that can be easily introduced.

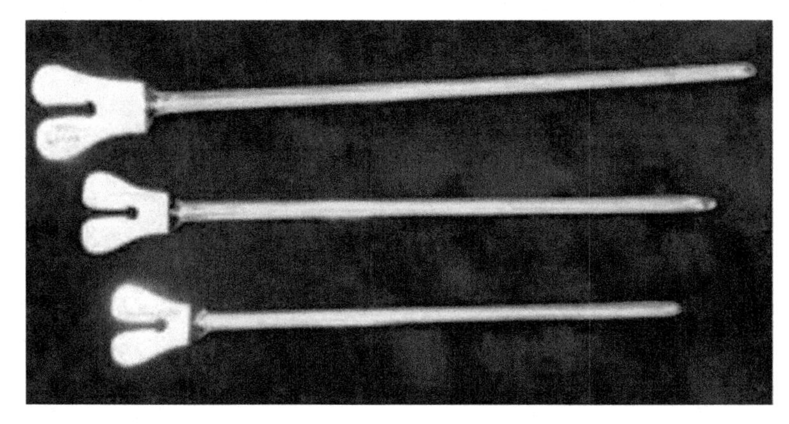

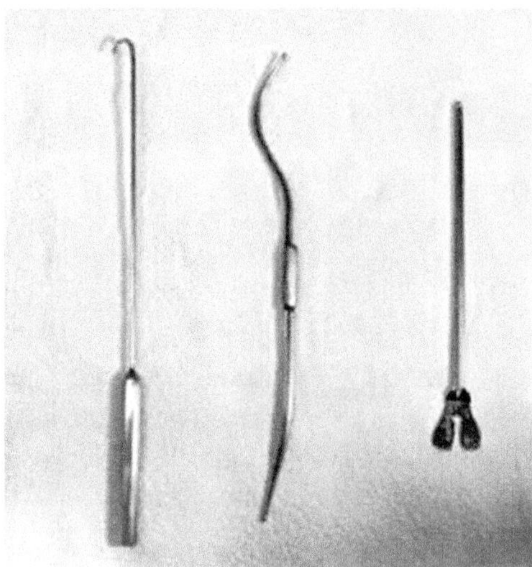

Fig 9.4 Choice of probes and directors

Pass it into the external opening and advance it towards a finger introduced into the canal (Fig 9.5).Next, lever the tip of the probe out of the canal (Fig 9.6). Be gentle. If you use force you may create a false passage and leave partly unrecognized, culprit track behind.. In most uncomplicated fistulae the probe will pass into the anal lumen at the dentate line.

Now incise on to the probe, thus uncovering the track. If the course of the track is complicated, incise over the probe piecemeal, and advance in short stages, uncovering its further course bit by bit as you progress. Should there be multiple external openings, first choose one closest to the anus. Ultimately you will have to probe all the openings and will find a connection from one to the other and eventually with the anal lumen. Constantly keep in mind a mental picture of the likely course of the tract, in accordance with Goodsall's rule and its corollary. Think two dimensionally: always be mindful that the track may course not only horizontally but also vertically.

Fig 9.5

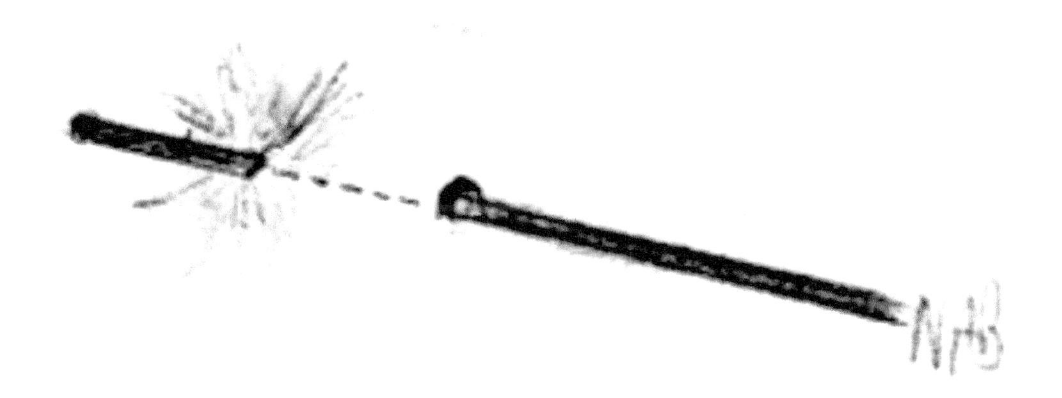

Fig 9.6 Bringing out the tip of the probe and positioning as as shown stabilises the tissues for the fistulotomy.

Should you encounter difficulty in finding the internal opening, frothy dye injection may help resolve the problem. A small quantity of frothy solution of Methylene blue in Hydrogen Peroxide diluted 1:10 and well shaken, then injected into the track through the external opening, will help identify the internal hole, when it is elusive. A temporary purse string suture around the external opening, pulled taught around the cannula, will prevent back leakage of the dyed solution and promote its flow onward and into the anal lumen. Use a small syringe e.g 5.0 ccs. Shake the diluted mixture vigorously to make it frothy.

Inject slowly with minimal pressure and watch for the site of appearance of the lightly colored bubbles.; don't flood the area.

Keep a constant watch on the anal lumen to identify the site of emergence of the dye. Inject slowly; over enthusiasm will flood and obscure the area. After the opening has been demonstrated, it must be confirmed with a hooked probe, passed retrograde, intraluminally. After the opening has been identified,, remove the canula and purse string and reintroduce a straight probe from the outside orifice in the certainty that you are not creating a false passage. Remove the speculum.

Once the probe enters the lumen, assisted by careful palpation, the course of the track is established as also the relationship of the fistula to the muscle ring. You may find it helpful to now carefully reinsert the speculum. Slowly open the blades. Gently lever the tip of the probe from within the anal lumen to be positioned outside and cushioned against perianal skin (Fig 9.6). This firmly stabilises the tissues within the operative field and the steps that follow can and should be precise and unhurried.

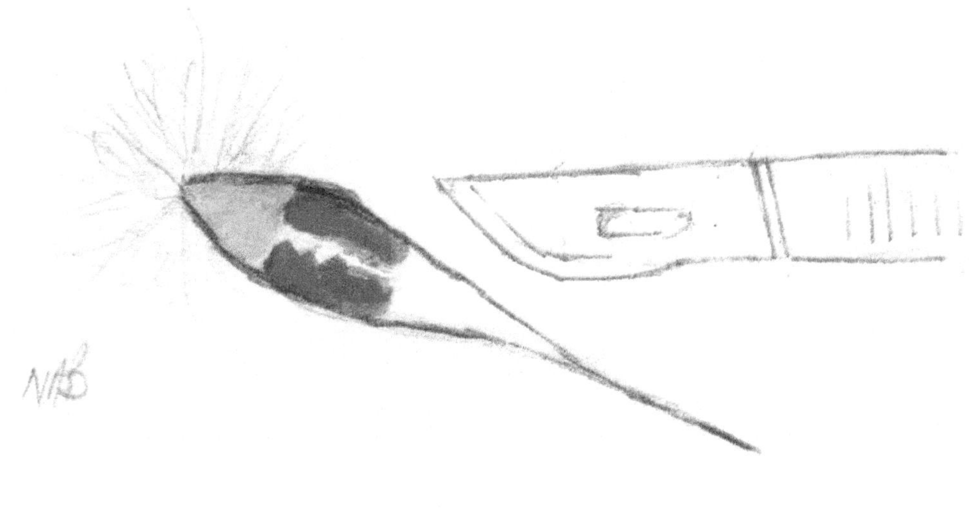

Fig 9.7

Depicting division of subcutaneous part of External sphincter to lay open the track. The main mass of the sphincter is left intact.

Make a shallow incision on to the probe using a scalpel. Once through the skin and anoderm, you have the choice to complete the division of tissues with the knife or diathermy (Bovie) point. If you opt for the latter, employ coagulation mode only , on a low setting.. Divide the superficial part of the external sphincter (Fig 9.7). Pause and carefully revaluate the internal sphincter before either laying it open or placing a seton.

Gently curette out the track , with especial attention to the intermuscular area, the space housing the intermuscular glands. It is unnecessarily mutilating to perform wide excisions of skin. A small representative tissue specimen is submitted for pathology. If there are multiple openings, each underlying connecting tract is first laid open over probes and the contents and walls are then curetted out.

A well padded dry dressing is applied.

The Seton

Where you have deemed complete surgical fistulotomy risky for causing incontinence, as an alternative, a length of non absorbable thread is left in the fistula for a variable time; from months up to a year. The most convenient material is a silk suture but some authors recommend fine rubber or plastic tubing. It can be placed via an aneurysm needle or a probe with a perforation

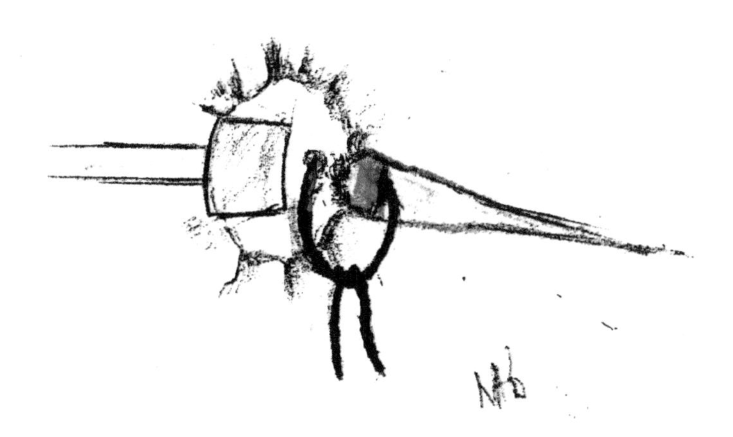

Fig 9.8 Here the seton is depicted loosely knotted

With few exceptions (usually from intermuscular abscess or after previous surgery) , the majority of fistulae open at the dentate line. They are named according to the depth and course pursued by the track itself in relation to the external sphincter and the ano rectal ring. Hence tracks as depicted in A, B and C above (low transmuscular or subcutaneous)) can safely be opened for their entire length. Those designated D, E and F (high transmuscular) endanger the uppermost portion of both the intermediate external sphincter and the internal sphincter and ring. with the risk of postoperative incontinence. These are more safely treated by placing a seton and staging any future surgery

There are then several objectives in using a seton. One is to mark and identify the course of the tract in relation to the anorectal ring, for revaluation at a later date under more conducive circumstances when the patient is conscious and can cooperate.

The second possibility is to allow the seton itself to gradually complete the fistulotomy by pressure necrosis. The rationale here is that prolonged presence of a foreign body will provoke an inflammatory response and the resulting fibrosis which will in turn ensure that the divided ends of the muscle do not retract, thus maintaining muscle control and function.

The third option is to maintain an open passage for prolonged local drainage, whilst allowing the neighboring tissues to heal, before its eventual removal.

Procedure

Probe the tract. It cannot be over emphasised that you must exercise care to avoid creating a false passage. Insert and open a bivalve speculum. If passage of the probe does not produce bleeding at its site of entry into the anal lumen you will be reassured that you have not created an iatrogenic false opening.

Curette out the space between the internal and external sphincters.

Next, thread a probe or aneurysm needle with a # 2 or 3 non absorbable tie (the seton). Knot the ligature tightly or loosely, according to your objective.

At outpatient follow up, gently tug on the seton whilst simultaneously performing digital examination of the anorectum. Instruct the patient to contract and / or bear down. This will provide reliable information about the sphincters and their relationship to the fistula. At this time a second operation, to complete the fistulotomy, may now be considered. If not, leave the seton for drainage until such time when you feel that it is no longer required. No anesthesia is needed to then remove it..

Recent Innovations.

Fibrin or matrix plug

The reported early successes of recent , minimally invasive treatments such as insertion of a fibrin or similar absorbable plug and preliminary experiments with acellular matrix point to a different approach for fistula in the future, which will hopefully obviate the division of normal muscle with its detrimental effect on continence; a giant leap forward in the treatment of fistula, particularly the high type. However the jury is still out and until these methods have stood the test of time, the traditional surgical approach will continue to serve. Presently, most reports indicate a significant failure to cure the fistula.

It is clear that whatever method is employed, it is crucial to remove necrotic debris, including the tissue from the culprit anal intermuscular glands. This is achieved by curettage of the intermuscular space and is probably the single most important step to achieve a cure.

Mucosal advancement flap

This delicate procedure is not recommended for the occasional anal surgeon. It is beyond the scope of this book and is included for information purposes only. In itself it does not address the problem of residual infected anal glandular tissue and curettage of the space will still be required.

CHAPTER 10

HYDRADENITIS SUPPURATIVA

This uncommon, minor but extremely disabling skin condition, affects the apocrine (sweat) glands of the perineum and groins and not infrequently, the axillae as well.

It is characterized by repeated abscess formation which discharges spontaneously and at first glance may be confused with anal fistula. There are however marked differences in appearance and behavior. There are characteristically extensive significant scarring and multiple small sinuses bilaterally, circumanally as well as away from the anus, on the perineum and thighs. There may be similar lesions in the axillae.

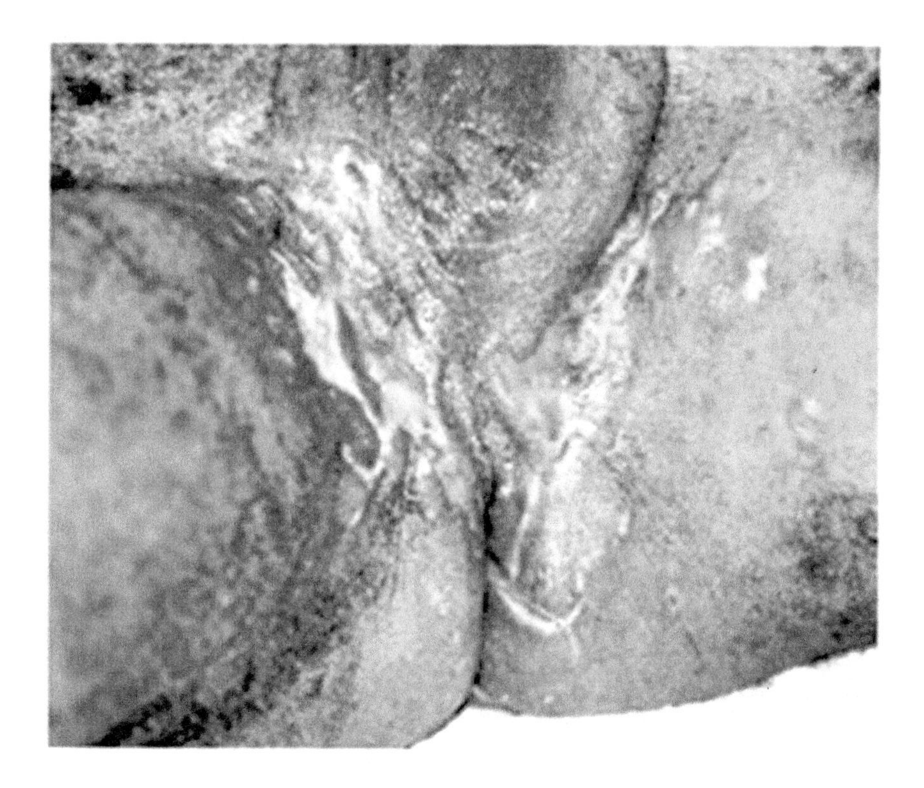

Fig 10.1

Probing with lacrimal probes will show that there are no tracks. Rectal examination will confirm the absence of underlying ano -rectal pathology.

Although local excision of affected areas may be of some benefit, the resulting scarring and disability mandates critical thought and planning as to how extensive any excision should be.

You may consider limited local (excision) of particularly troublesome areas, leaving them to granulate and heal by second intention, or, where appropriate, secondary split skin grafting at a later date, when local conditions permit. Skin grafting early on, in the presence of active infection and pus, is problematic.

Once this problem is recognized and correctly diagnosed, you should seek advice from a dermatologist and you may also consider working in collaboration with a plastic surgeon.

CHAPTER 11

CONDYLOMA ACCUMINATA

This highly infectious, papilloma viral disease, is sexually transmitted. Hence, not unexpectedly, it is prevalent, although not exclusive to, the homosexual population. It attacks the superficial layers of squamous epithelium predominantly, in warm, moist areas such as the peri- anum, the perineum and the pubic areas. Isolated lesions may occur on skin well removed, e.g. penile shaft, lower abdomen and thighs. It is not seen above the dentate line . Excrescences may be fragile and bleed from slight trauma from underware and therefore may be uncomfortable, or they may be asymptomatic.

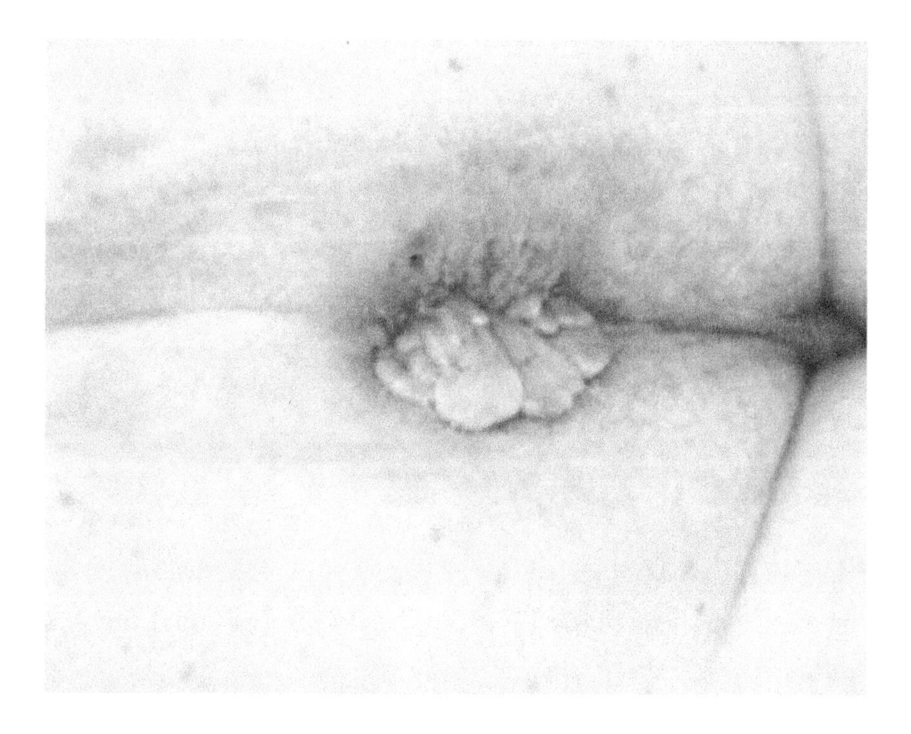

Fig 11.1

The author, once only during a long career, encountered a single lesion, just above the dentate line.

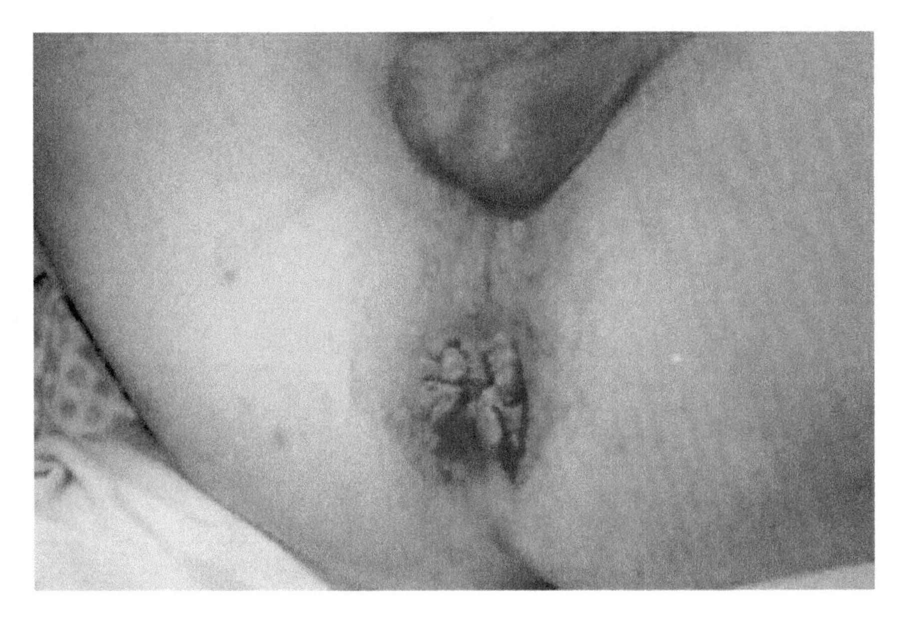

Fig 11.2

Pathology

The lesions are moist and succulent and may be scattered, isolated or profuse. A rare form of Giant Condylomas (Boeschke- Loewenstein) may have malignant potential.

Treatment

Wide excision: Although this is commonly practised, in the author's opinion it should be avoided. It is mutilating and disabling and the likelihood of new lesions is high.

Fulguration and curettage:

There is no limit to the size of the area which can be treated in one session.

Wear magnifying loupes to ensure that small lesions are not missed.

Protect yourself and the operating room from inhaling vapor containing the live virus. There have been reports of the virus producing condylomas on the vocal chords. Therefore wear a double mask and keep continuous suction present right at the site to immediately siphon away fumes.

Set the Bovie (a.k.a diathermy machine) on coagulation, at a medium/ low setting. Various lesions are lightly "sizzled". They are succulent and they rapidly "boil". Next, lightly curette away the dessicated lesions. Bleeding is minimal. The improved appearance is striking; after completing curettage, the skin appears to be only superficially and minimally excoriated but otherwise intact.

Do not overlook the intra anal lesions and scattered lesions further afield.

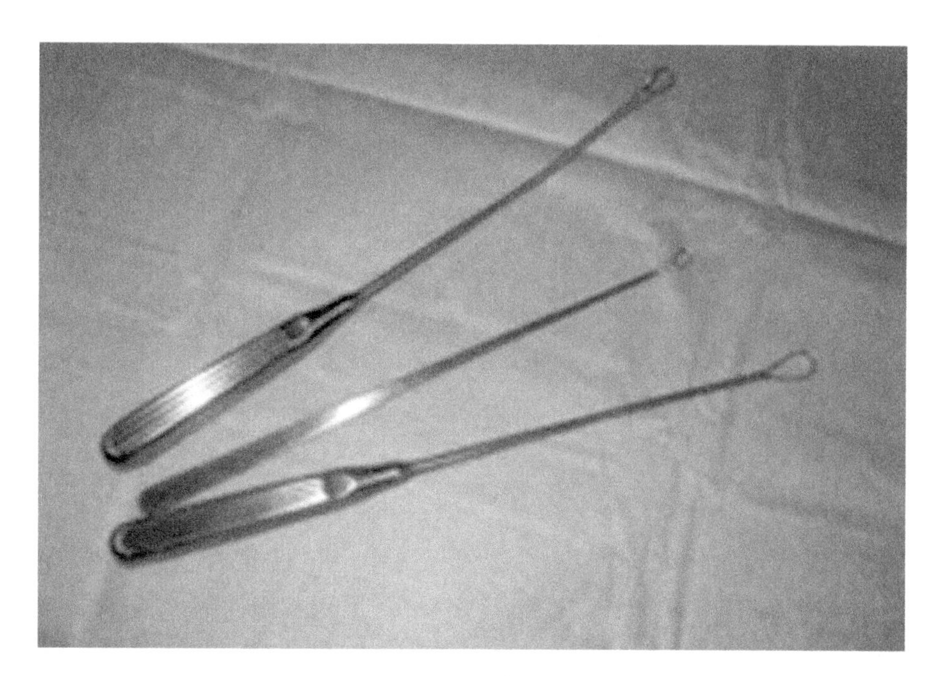

Fig 11.3 Assortment of Uterine Curettes

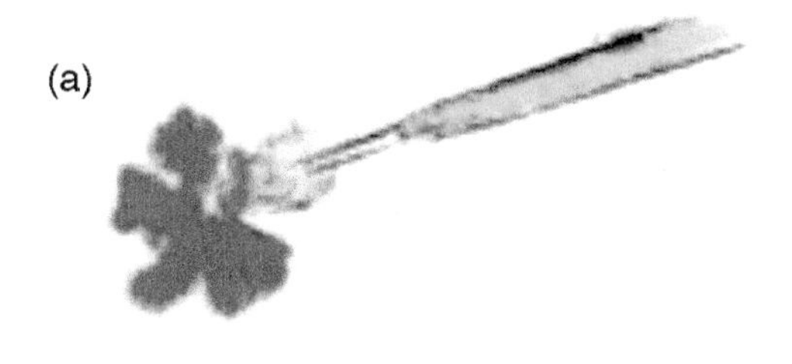

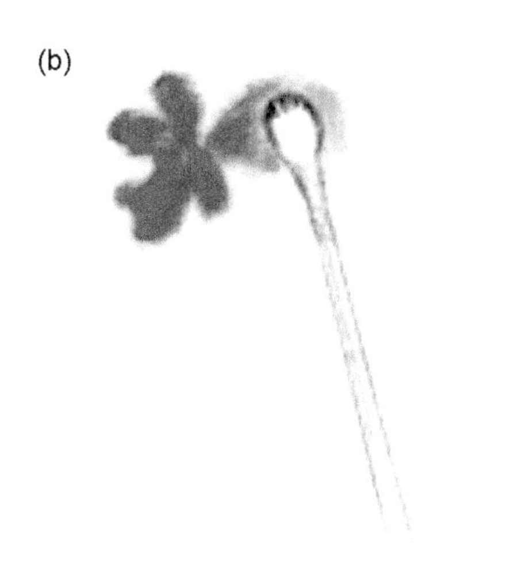

Fig 11.4
(a)First "sizzle" (b)Next curette away

Smother with a generous layer of 5% Efudex (5% fluoro uracil) cream liberally applied to the entire raw area as well as intra anally. Apply a well padded absorbent dry dressing.

Masses of tissue are available to submit for histopathology.

There is minimal pain and virtually no post operative disability.

At ten day follow up the area should be healing well. Apply another dose of Efudex cream. The etiology of the disease is clearly explained to the patient, as also the risks of recurrence as well as other diseases if continuing his/ her lifestyle.The patient is followed up indefinitely insofar as there is compliance. The etiological role of the patient's lifestyle and necessary reform are repeatedly explained.

Healing is rapid and the final appearance is quite normal, without scarring and without any skin loss. In a considerable experience using this simple method, the author found the results to be gratifying with zero mutilation and zero recurrence. Should this however occur, the procedure could be repeated as necessary.

CHAPTER 12

PILONIDAL SINUS

In reality this is not an ano rectal disease but due to it's proximity to the anus and the possibility of confusion with anal fistula, it has come to be placed in that category. It is a minor condition that can produce major disability. As the name ("Nest of Hairs") implies, it is characterised by the presence of hairs in a subcutaneous chronic abscess cavity, which is located in the distal natal cleft. Do not be fooled by an abscess pointing lateral to the midline. Close inspection will show one or several small openings in or off the midline, which may or may not manifest drainage (Fig 12.1).

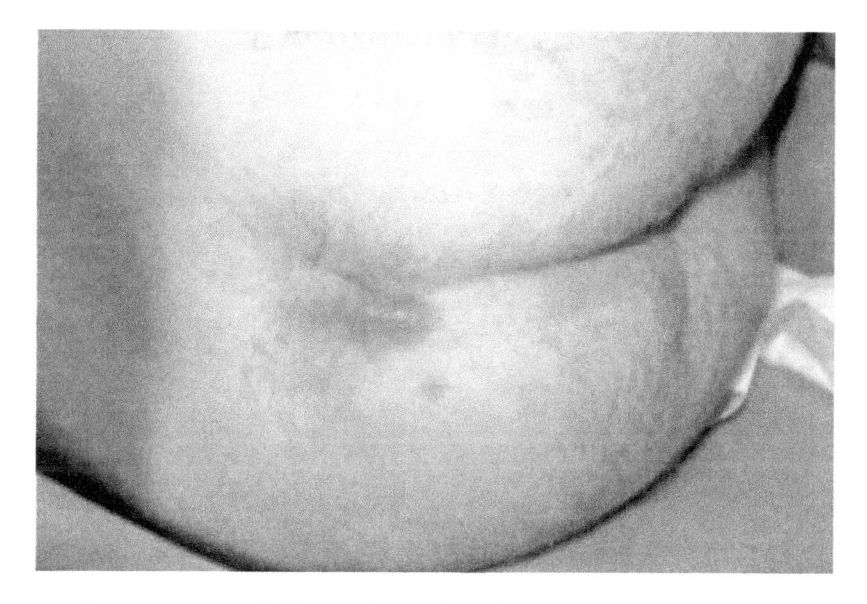

Fig 12.1 Abscess pointing off the midline. Showing multiple sinuses in both the midline and laterally. They will all communicate with one common central cavity.

Although patients tend to be hersuit, this is not invariable. The debate regarding acquired versus congenital etiology is not fully resolved but for several decades and currently, the majority of views favor the former.

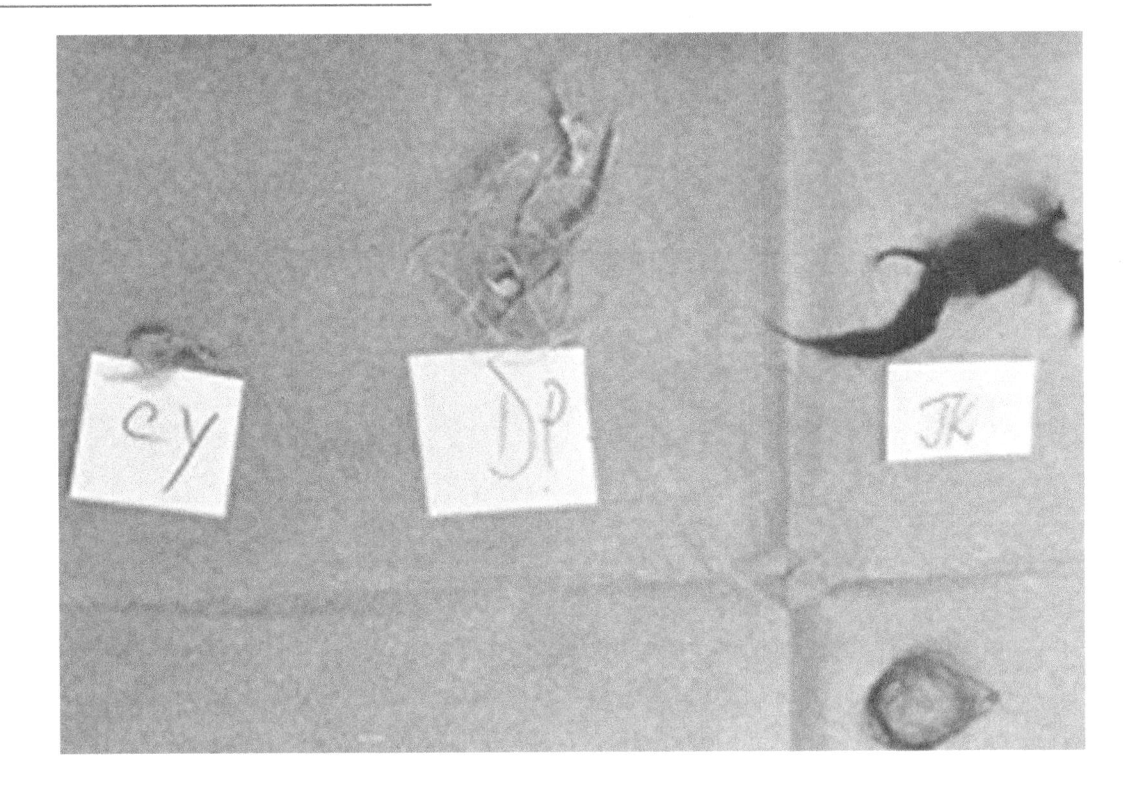

Fig12.2 An assortment of hairs from different patients. Hairs vary in quantity, clusters, consistency and disposition. Their color and consistency may be different from neighborhood hair. This casts some doubt on the current popular theory that the disease is acquired by penetration from local sources.

The quantity of hairs may range from large masses to a single strand . Furthermore the hairs may not resemble those of the adjacent skin.

Misconceptions about the pathology, in addition to overly aggressive and mutilating surgery, have resulted in inestimable loss of work hours and disability as also the apparent belief that this minor condition is inexplicably difficult to cure.

Conservative Operation

The procedure described hereunder is equally applicable in the presence of an abscess or a quiescent lesion. Note: It can also be used for recurrences.

Local anesthesia can be used in some circumstances but general anesthesia is preferred.

Position the patient prone. The table is “broken”.

No matter where the discharging sinus, it is mandatory that the midline pathology be openly displayed and eradicated.

Good lighting is essential.

Following meticulous shaving of the area, remove all loose hairs with adhesive tape. Paint the buttocks with Tincture of Benzoin Compound and allow to dry. Apply adhesive strips, pull laterally and secure them to the side of the operating table. This flattens the natal cleft and achieves excellent exposure.

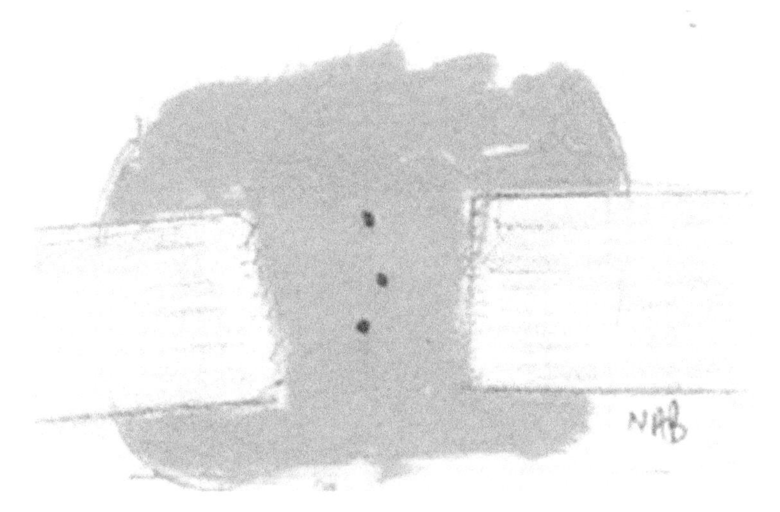

Fig 12.3 The strapping exerts lateral pull and flattens out the natal cleft.

Start probing with lacrimal probes. If tracts permit, you may decide to switch to regular probes. Lacrimal duct probes are in place.

The red dotted line indicates the incisions. The sinuses all lead into one common underlying chronic abscess cavity.

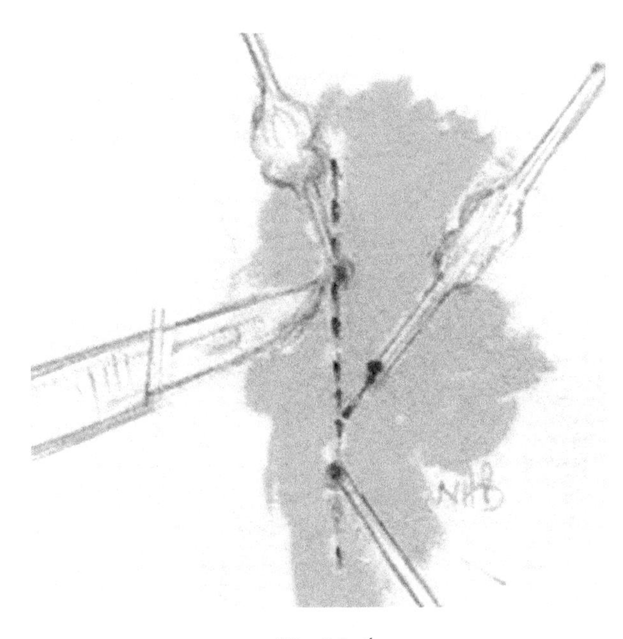

Fig.12.4

Incise with a #15 blade onto each probe. Each opening will lead directly to the same common underlying cavity which is fully opened up (Fig 12.5). Hairs are wiped out with a small gauze sponge and meticulously removed from the operating table.

Next curette out the floor and side walls of the cavity (Fig 12.6). Excise a narrow (1.00 mms or less) rim of skin around the edges of the wound. Send tissue for pathology.

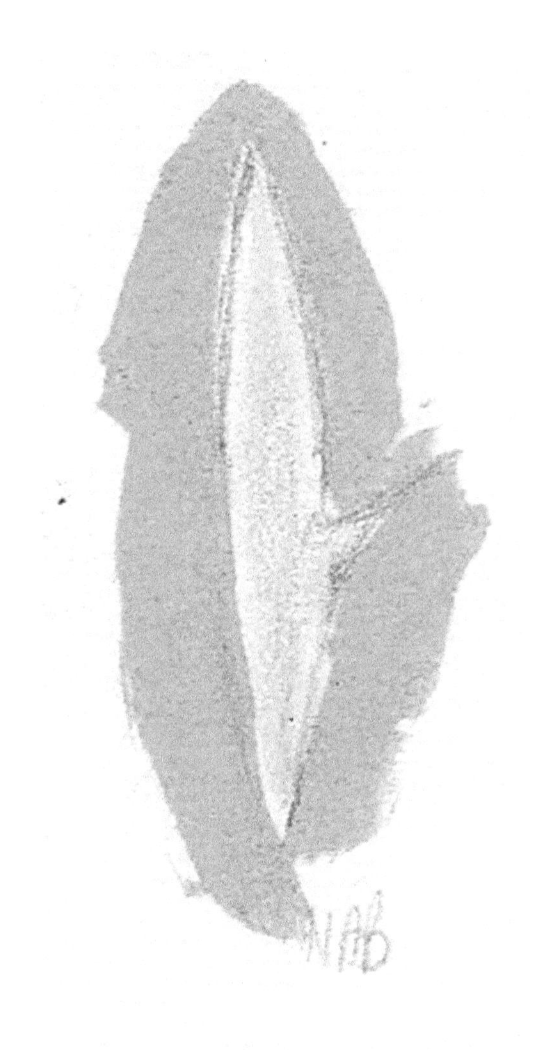

Fig. 12.5

References:

Blumberg N A. Pilonidal Sinus treated by Conservative surgery and the local application of Phenol S Afr J Surg 1978; 16:245

Blumberg N A. Pilonidal Sinus. An old problem revisited. Surgical Rounds June 2001 pp 307-316

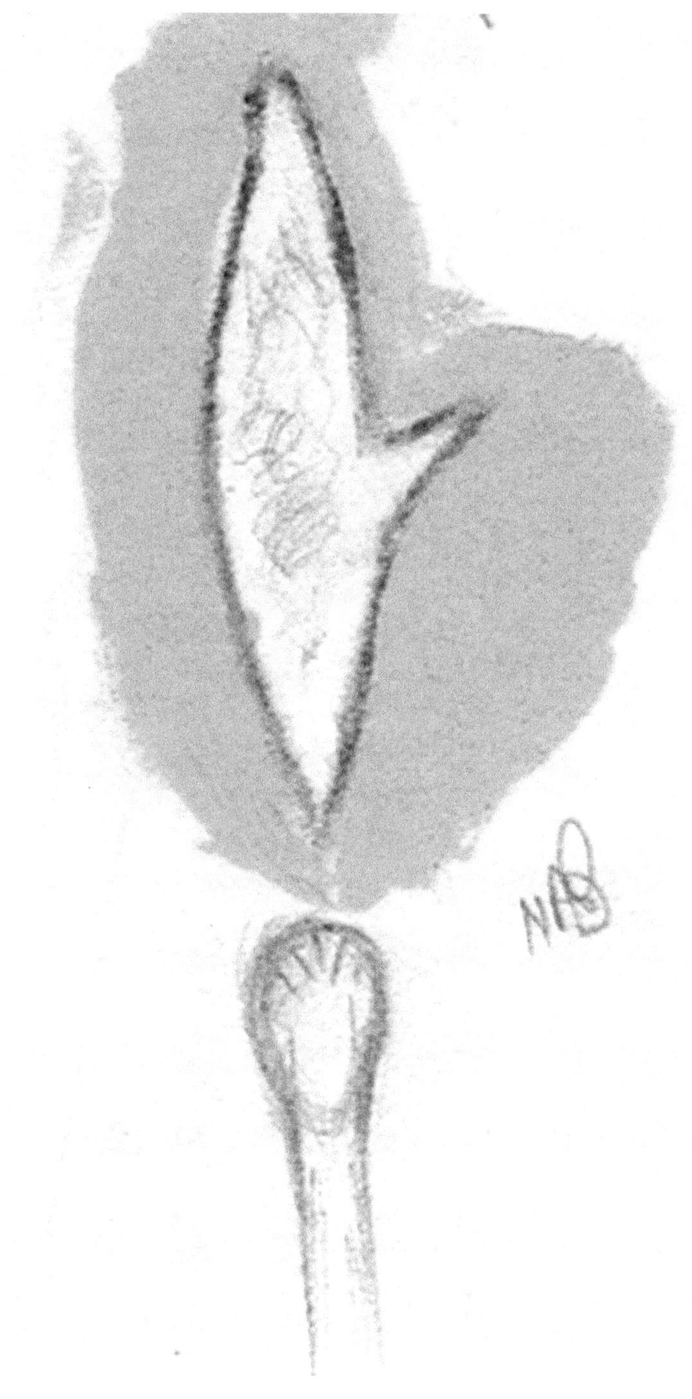

Fig. 12.6

Now smear the skin around the site with a thick protective layer of petroleum jelly (Vaseline). Cover the anus with a sponge to keep Phenol away. Next, soak the cavity with 10% Phenol applied on a small sponge. The tissues immediately turn white. Wipe it out after a ten or so seconds. Be careful to keep the phenol away from the scrotum. Should this however occur, there will result superficial chemical burns, which will heal without permanent damage.

Leave the wound open. Apply a dry dressing.

There are no special after treatment nor restrictions except to wipe out the cavity with a Phenol soaked sponge. This is painless. Some hairs, previously undetected, may surface but you must first and at all times protect the skin with petroleum jelly against Phenol burns. The patient is followed every two weeks in the office. If there is local regrowth of hair, the area around the wound is carefully shaved; the utmost care is taken to keep loose hairs out of the wound. To this end, place a small cotton wool pad inside the cavity whilst shaving and remove when through.

The wound closes rapidly. Periodic shaving is practised until the wound is completely healed and epithelialised. Ideally, the patient should keep the area hair free permanently.

There is no postoperative disability and the patient can return to work after a day or two.

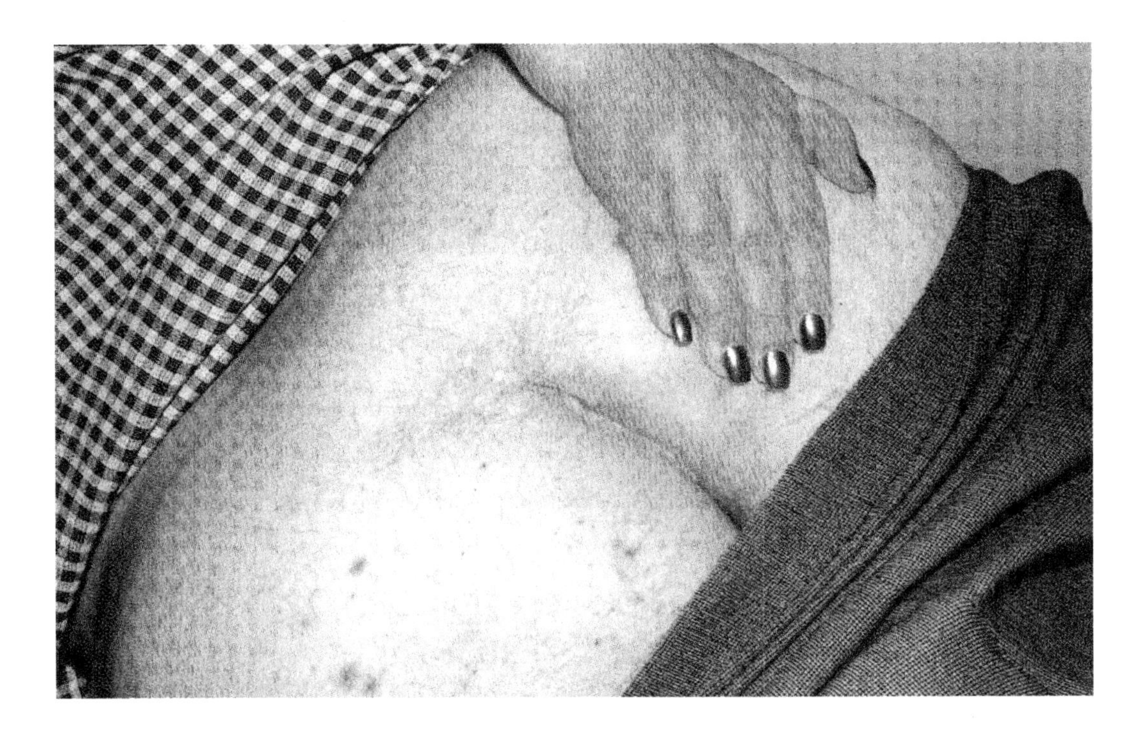

Fig. 12.7

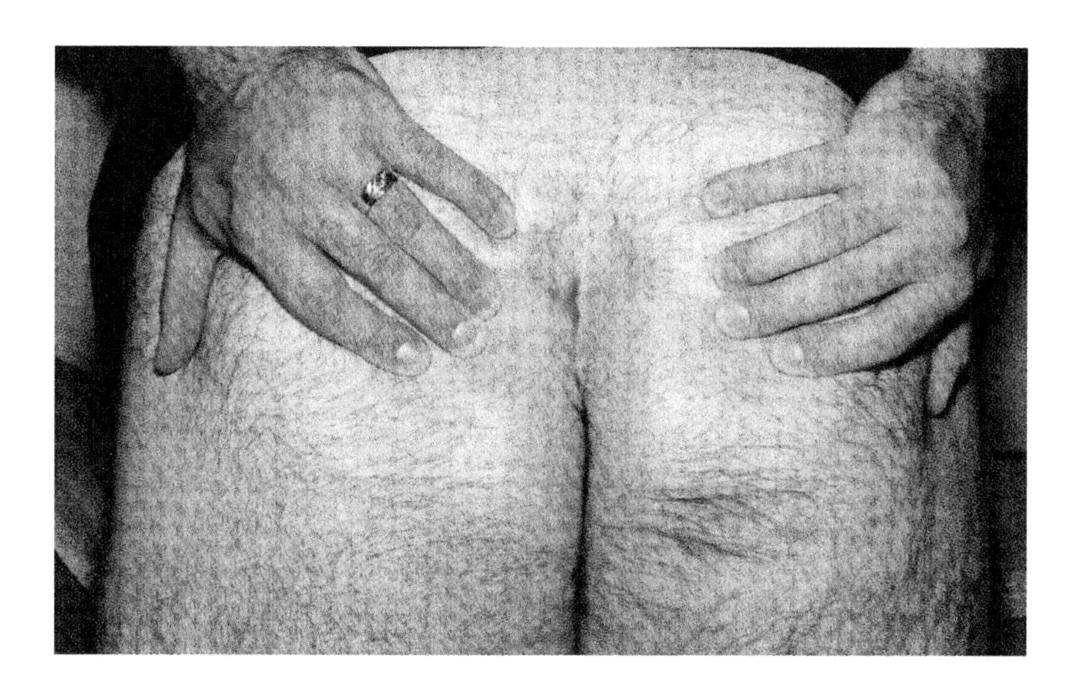

Fig 12.8

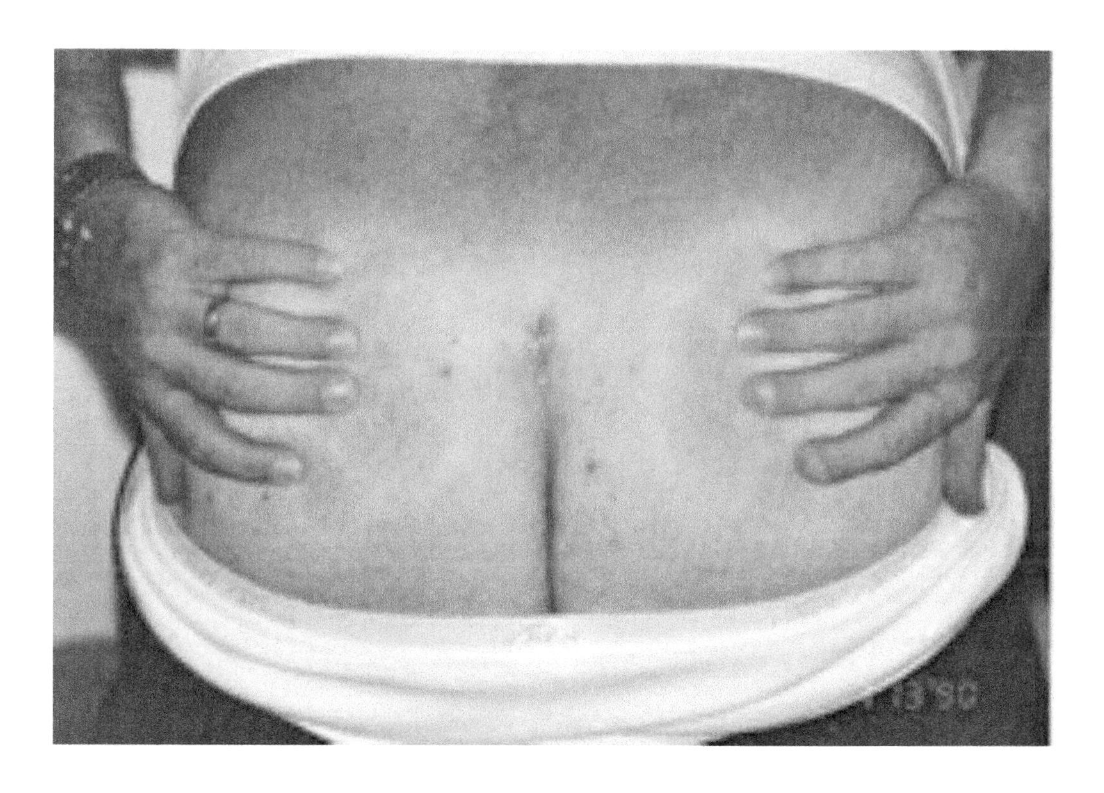

Fig 12.9

Three patients, each exhibit healing with a postoperative fine linear scar.

CHAPTER 13

PRURITIS ANI

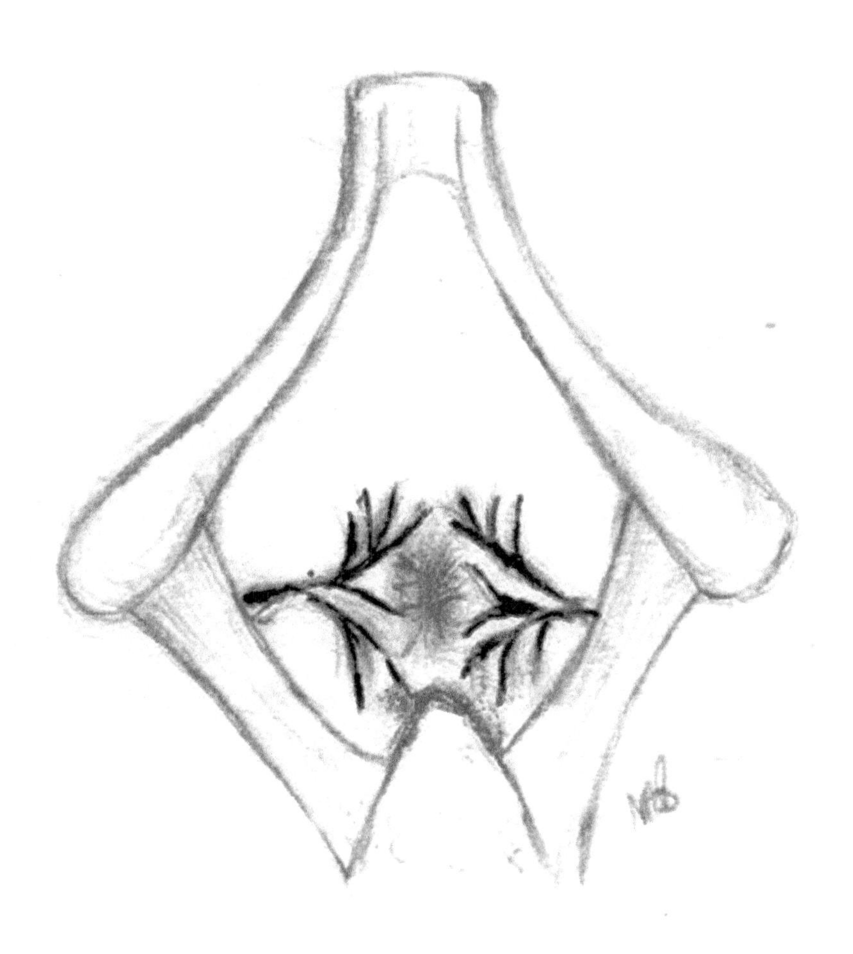

Fig 13.1 This schematic depicts the rich sensory innervation of the perineum behind the urogenital triangle: the inferior haemorrhoidal nerves coming in from laterally (Pudendal nerve within Alcock's Canal).

Pruritis Ani is a common condition. Common underlying causes are poor local hygiene, bowel parasites e.g threadworm, excessive sweating with poorly maintained underwear, diabetes, fungal infection, allergy to the washing materials used in the underware, mucoid leak from third and fourth degree hemorrhoids.

The degree of disability varies widely from minor to gross. The patient may be miserable beyond expectation from local findings, even in some cases harboring suicidal thoughts. In severe cases there may be severe excoriation from scratching and iatrogenic ulceration. There develops a vicious cycle from scratching to more excoriation and more scratching. Patients have switched from one ointment to another, have tried talcs, shaving and drying agents, often to no avail.

The first step in treatment should be meticulous hygiene of the area. Immediately following bowel activity the area should be thoroughly washed (no soap) and thereafter kept dry. This may have to be done periodically during the day even without bowel movements and before bedtime. A home -prepared Sitz bath of several teaspoonfuls of ordinary table salt dissolved in a bowl of tap water employed twice per day or as needed, will soothe and provide significant relief of symptoms.10

Underwear must be clean and dry. It must be well rinsed after laundering and free of all soaps and detergents.

It is axiomatic that the treatment should be directed against any obvious cause. However, there exists a small cadre of patients in whom extensive, repeated investigation will reveal no underlying pathology. Their disease can appropriately be labelled "idiopathic". Long term nerve blockage, the procedure described hereunder, is a desperate one to be considered only after all else has failed; it is rarely called for. It can be highly successful and also may be repeated if there is recurrence of symptoms. It is not however without risk for morbidity e.g. skin necrosis and abscess. These are fortunately not serious, but patients must be so informed. You should have a witnessed signed consent document to this effect, within your patient record.

These will invariably eventually resolve, leaving no permanent disability.

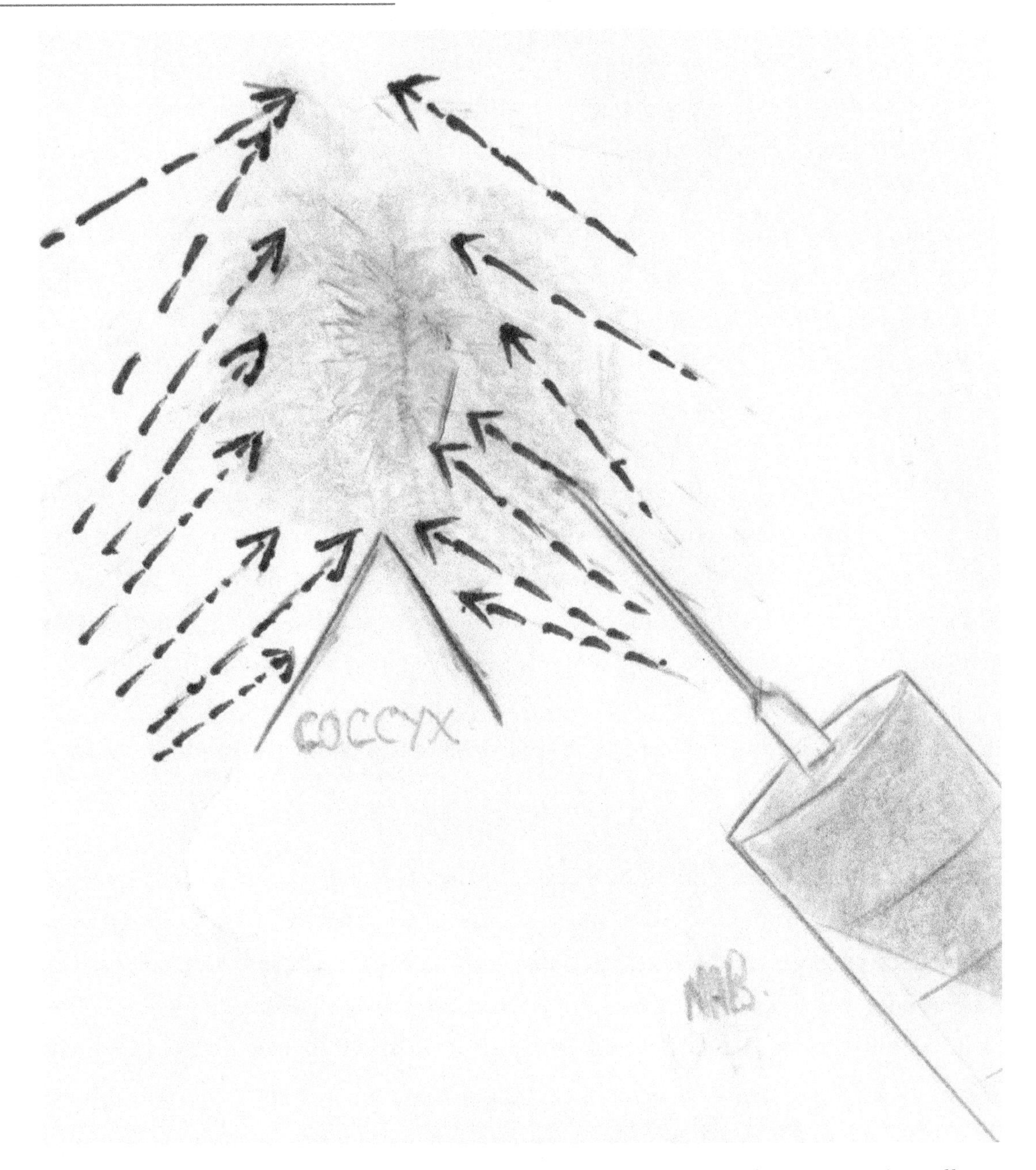

Fig 13.2 Depicting subcutaneous infiltration of 70% alcohol.. Multiple punctures (usually 5 to 10) are necessary to ensure complete circumanal blockage. Be especially careful not to penetrate into the anal canal lumen or the penile bulb anteriorly. An index finger inserted periodically into the anal canal will assist in placement of the needle and help protect.

Procedure

General Anesthesia is mandatory. The patient is positioned prone A subcutaneous injection of 50 to 70 ccs of 70% alcohol is performed as per the illustration. Five ccs of the solution are injected into each of the ten or twelve puncture sites. "Fanning it out" ensures that no area is overlooked. "

Ideally, to reduce the high risk of secondary infection, the skin would be clean and non- ulcerated but this will probably not be not be the case in those patients where you have to resort to this radical treatment.

Inject slowly and watch for blanching and edema of skin. If this occurs the needle is placed too superficial and needs repositioning. Do not inject intradermally. Avoid piercing the anal canal wall; guide by placing an index finger within. In addition, in male patients, avoid the area close to the urethral bulb With each repositioning, first withdraw the plunger to ensure that the needle tip has not entered a vein- an accidental intravenous injection of this medication could be disastrous.

Relief of disabling itching may be dramatic and cure will likely be permanent. If not, the procedure can be repeated as needed.

Complications

There is a high risk of infection locally, especially where the skin is badly scratched and/or infected. There may also occur patches of skin necrosis, from an incorrectly placed injection and therefore the injection must given very slowly. Watch the skin closely for blanching or edema, signifying that the injection is intra cutaneous. In that case, immediately withdraw the needle and reposition it subcutaneously. Antibiotics should be taken orally

Local abscess may develop. This will likely rupture and discharge spontaneously without the need for surgical intervention.

CHAPTER 14

HYPERTROPHIC ANAL PAPILLA

Anal papilla may be solitary or multiple. These are commonly thought to be remnants of the proctodeal membrane. They may be exquisitely sensitive.

They may prolapse and protrude through the anal orifice and are sometimes incriminated as a cause of anal fissure. This hard-to-imagine scenario is said to occur where the downward pull, from propulsive bowel movement, tugs at the mucosa and tears it. The main indications for removal are prolapse outside the anal canal with its effect on local hygiene, mucous leakage and in order 'to submit for histopathologic evaluation.

The papilla is smooth and firm and has a clear benign appearance. The base is avascular and it is usually not necessary to ligate it.

Procedure: Crush the base with an artery forceps (hemostat) prior to excision. If you so desire, tie it off with a fine non absorbable ligature. This will come away spontaneously and be passed after a few days.

The operation is easily performed via a proctoscope. If it is large and prolapsed (Fig 14.1) the papilla can be removed extra-anally, usually under Local Anaesthesia but it may require General Anaesthesia.

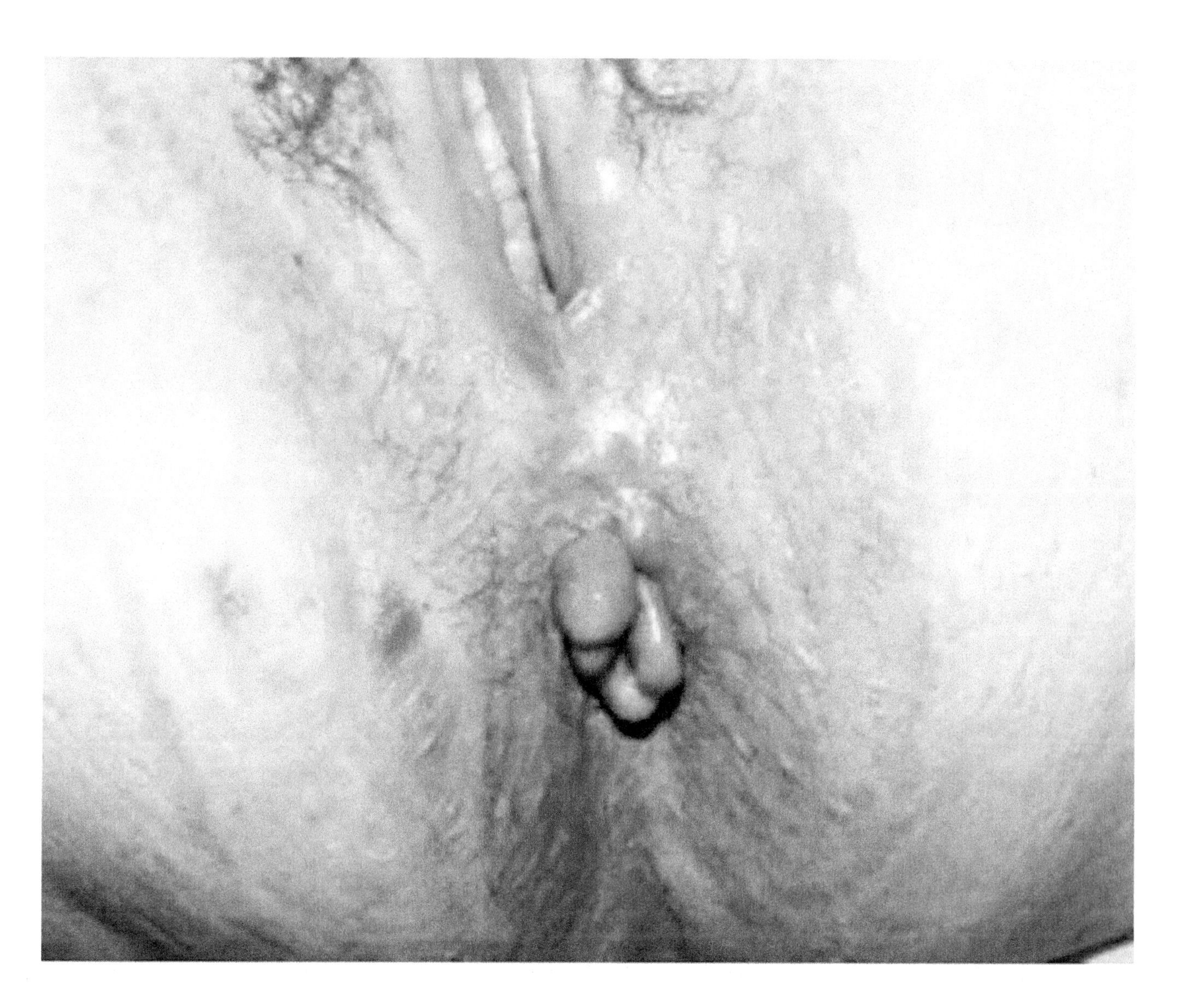

Fig 14.1 Multiple protruding anal papillae

CHAPTER 15

ANAL INCONTINENCE, PROLAPSE IN THE ELDERLY

There are numerous causes of this disabling and embarrassing condition which may vary in degree from minor, e.g. flatus and some liquid stool, to gross incontinence. The most significant cause is trauma, mostly obstetrical, as seen commonly in third world populations. Another iatrogenic cause is sphincter damage from surgery, such as for complicated anal fistula. However, the quite commonly encountered incontinence in the elderly, is a result of natural neuro- myenteric attrition. This is invariably combined with a degree of ano rectal prolapse. Here the surgery is within the ambit of the General Surgeon.

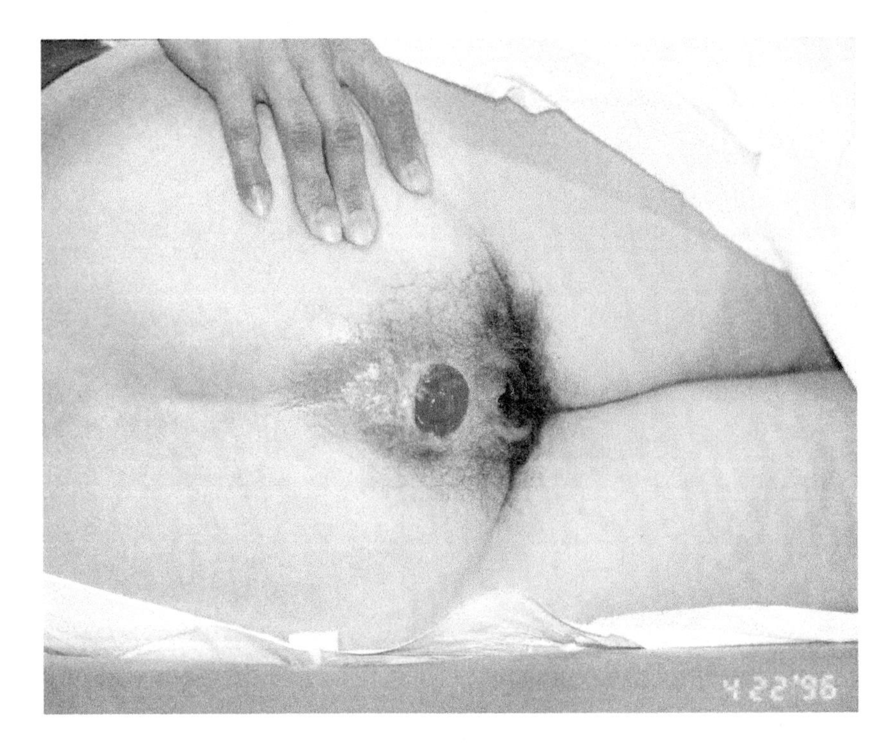

Fig 15.1

Inspection will show a patulous anus. On digital examination there will be absence of muscle tone and inability to voluntarily contract the anal muscles around the examining finger.

Highly complicated circumrectal muscle reconstruction procedures such as the Delorme procedure, are beyond the scope of this book. These are the province of the super specialist, not the occasional operator. One only simple procedure, which has stood the test of time, is described.This minor operation can be effective in the elderly and it is easily performed under local anesthesia.

The traditional material used was fine steel suture material. Although effective, it has many potential problems: wire fracture, patient discomfort from the knot or feeling the ends of a suboptimally placed wire ring. Infection can occur. Any such complication is readily dealt with by removal of the suture and its replacement if and as often as necessary. For many years the author employed heavy Nylon sutures with satisfactory results.

Thiersch Procedure

The patient is positioned in lithotomy if that can be tolerated. Although not ideal, it can also be performed in lateral decubitus should that be necessary.

1% lignocaine is infiltrated circumferentially around the anal orifice.

Make two separate 0.75 cm incisions at the anal verge, one at 12.00 and one at 6.00 o'clock.

The material can be placed using an aneurysm needle, a Doyen's needle or even a fine hemostat via the posterior incision. It is immaterial whether you move from right to left or left to right; follow your own preference.

The suture is inserted deep subcutaneously and within the substance of the superficial external sphincter, Keep an index finger inside the canal to ensure appropriate positioning, to avoid over narrowing and also to protect against creating a false passage into the anal canal lumen.

Fig. 15.2

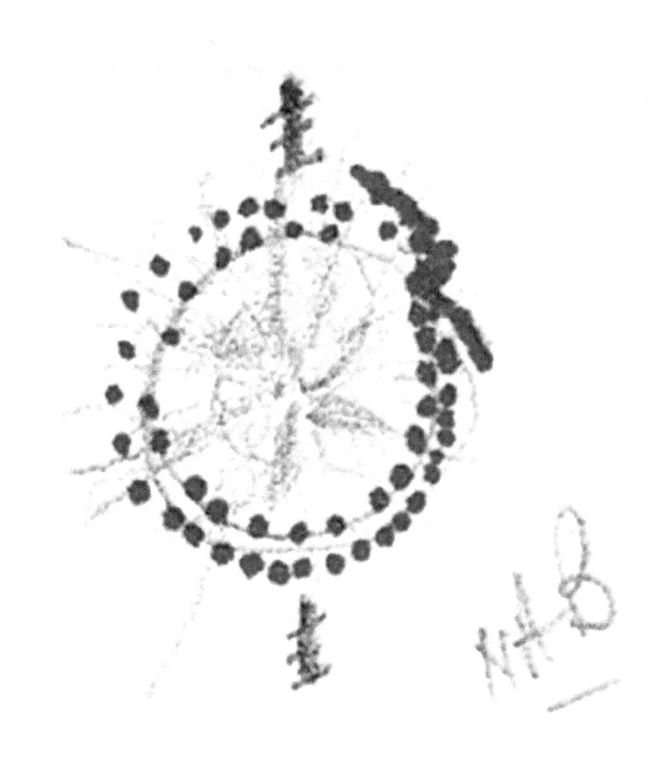

Fig. 15.3

Pass the suture circumferentially around and parallel to the anal verge , then out through the anterior incision.

Next repeat the placement on the second side, starting at 12 o'clock. Bring the suture out at six o'clock, where the two ends can later be securely knotted.

Double the suture by a second running, before knotting.

Pull the suture sufficiently tight to leave a 1 finger lumen. Repeatedly, digitally and visually confirm placement, burying of the knot and the avoidance of anodermal puncture.

After multiple reef knots, to ensure against unravelling, the ends of the suture are deeply placed within the superficial external sphincter substance, using a small round bodied needle.

Repair the skin incisions with a few 4/0 Vicryl or Dexon interrupted sutures placed subcutaneously.

Additional Treatments

Pelvic Floor Exercises

Although this is theoretically a useful addition, in reality it is not easy to achieve The patients are usually elderly and cannot be relied on continue to practice these exercises for years on end let alone permanently.

Attempt to solidify quality and consistency of stool by oral fibre preparations such as Metamucil.

CHAPTER 16

VILLOUS TUMOR

Rectal bleeding and fluid/ mucoid discharge always mandate complete rectal and colonic evaluation. Even where the culprit tumor has been found in the rectum, additional tumors or polyps may coexist elsewhere in the colon. Total colonoscopy is thus mandatory in every case.

Villous tumor, an essentially benign lesion, has an enormous surface area due to branching and rebranching of the multiple fronds ad infinitum. Hence it may harbor obscured areas of malignancy which escape ready detection. These tumors may be so soft as to be be impalpable and digital examination may not readily detect areas of induration within. Because of the enormous surface area, there may occur loss of large quantities of fluid and electrolytes, especially potassium.

There may be an unsuspectedly large tumor on either a relatively small or a large base of attachment. Although complete removal can usually be accomplished in one session, large size may lead to two or more operations to achieve complete removal. Where the base is large, removal should include a small margin, 2 to 4.0 mms all around, of normal mucosa.

Procedure

Under general anaesthesia, insert a double or three bladed retractor.

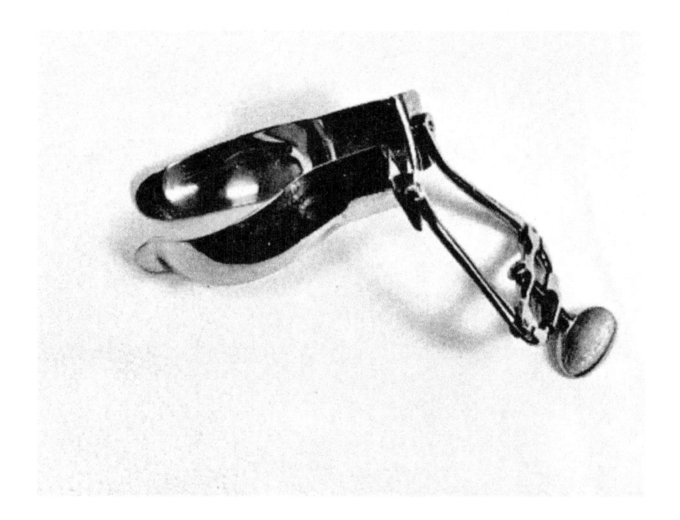

Fig 16.1 Parks Mark 1 retractor

Where the tumor comes off a small stalk, removal is usually quite simple.It is the large lesion with a large base of attachment that may present difficulty.

It is usually easier to commence dissection inferiorly, as depicted in (a) and (b) below.

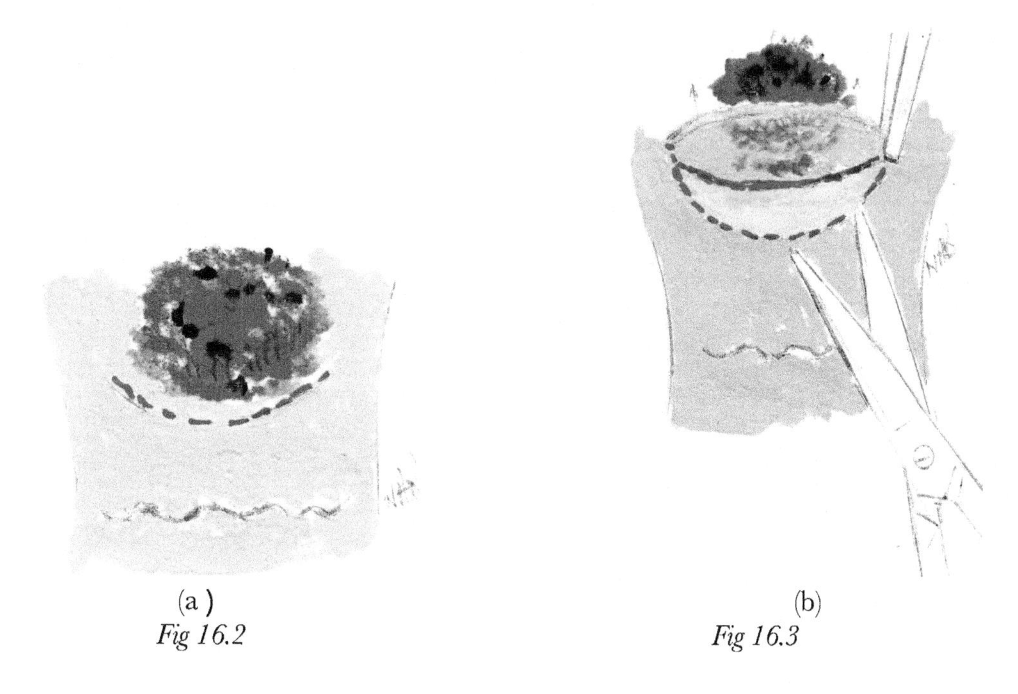

(a) *Fig 16.2*

(b) *Fig 16.3*

Infiltrate submucosally the proposed areas of dissection as you progress, with 1;100000 epinephrine in saline. This raises the mucosa and readily defines the plane of dissection. Exercise the utmost care to not penetrate through the bowel wall.

Periodically mark the edges with temporary fine silk stay sutures. These simplify the later repair, at which time they are removed.

Scissors dissection is preferred. As the dissection progresses and the specimen becomes more clearly defined, grasp with Judd Allis clamps. Constantly reapply and reposition as needed.

A double box jointed needle holder (c) affords easy suturing without obscuring the field. Repair the mucosal defect, horizontally if possible (d), with interrupted heavy absorbable sutures, securely knotted multiple times. Each suture should also bite into and include the internal sphincter muscle mass. Apposition should be tension-free. Leave the ends of the sutures half to one inch long to ensure against unravelling.

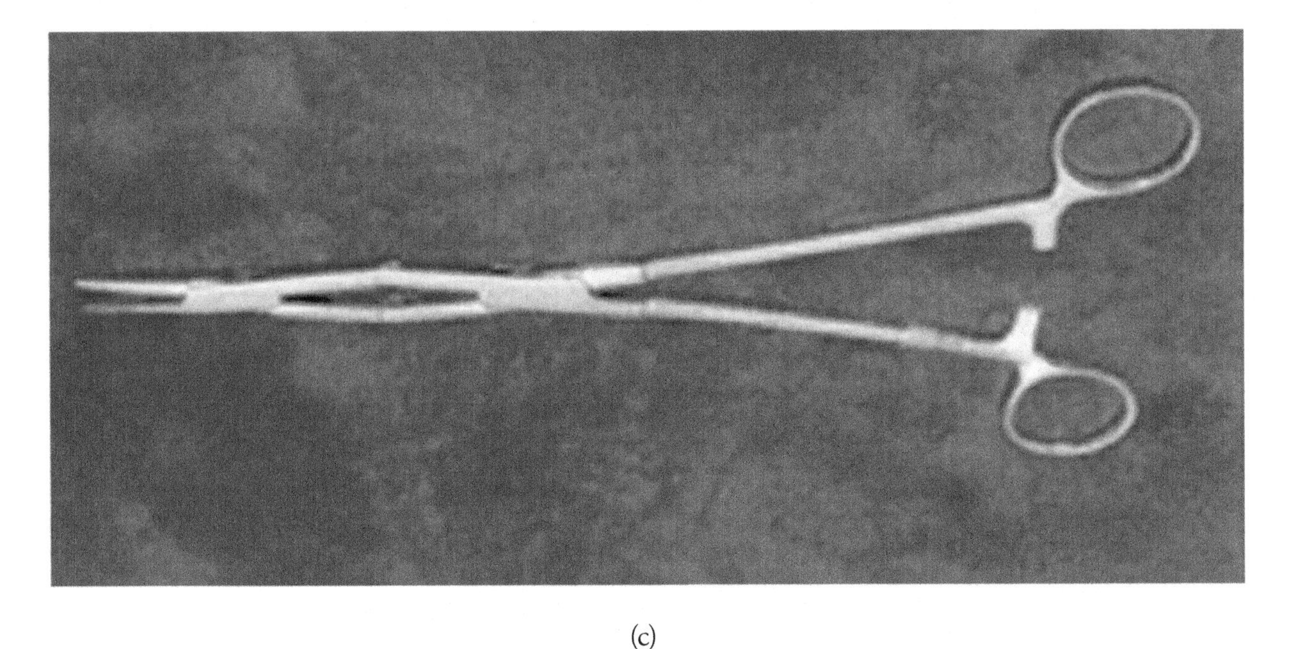

(c)

Fig 16.4 Double boxed needle holder

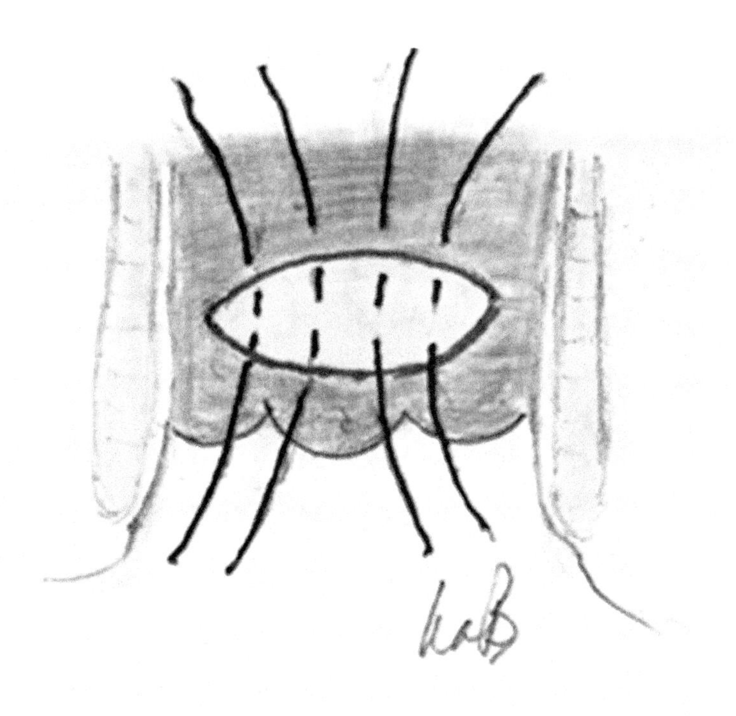

Fig 16.5. Sutures are placed vertically. Placing them horizontally may be tempting since it would be easier, but that could produce anal canal narrowing.

If the lesion is so large as to obscure the dissection, you may start by removing it piecemeal to enhance visibility. On completion of the dissection you may find that there is a relatively small stalk, so that final repair should be easy. However, in the rare situation of a massive villous tumor, so large that it completely encircles the bowel wall, local excision may be impossible. This would then require a formal resection of the rectum with anastomosis or possibly even an abdomino- perineal resection, depending on the level and other local factors. This circumstance is beyond the scope of this book and here readers are referred to standard texts.

CHAPTER 17

MISCELLANEOUS

Dealing with severe constipation (obstipation)

Even though it may not be etiological, there may coexist a close relationship between many of the diseases in this book and constipation, which may be severe. Pain and cough medication containing Codein may be a potent culprit.

Treatment of idiopathic constipation may cover multiple facets:

Diet: Bulk and high roughage intake are essential.

Stool Softeners as and where indicated. These are over the counter products. They must be taken with significant amounts of liquid.

Glycerine suppository for impaction

Mineral Oil should not be used as a regular laxative. Whereas it is satisfactory for occasional constipation, it has a tendency to leak or remain on the the perianal skin with prolonged use. In the presence of skin tags or other local pathology this may in itself create problems with personal hygiene.

Where there is obstipation or where the sensation demanding defaecation is present but there is difficulty with expulsion of impacted stool, adopting the position illustrated can be of tremendous assistance. This introduces a significant degree of additional muscle power, which it transfers to the abdominal muscles.

Fig. 17.1

It effectively combines several forces: Mechanical assistance to the already contracting abdominal muscles, from the shoulders and upper extremities in addition to the thigh muscles. Whilst on the commode, the arms are pulled tight against a leg with the knee in flexion. The hands may be locked if needed.

Take three or four deep breaths (this splints the diaphragm) and pull hard. Pause, then repeat as needed until the bolus is expelled. Especially in older patients, who show muscle attrition and weakness, adopting this position and manoeuvre will usually overcome even in the most stubborn cases.

FURTHER READING

Regarding Internal Sphincterotomy for Anal Fissure. Norman A Blumberg 1998 Dis Colon and Rectum 41.8 p1071- 1072

Anorectal Surgery. E S R Hughes and A M Cuthbertson Chapman and Hall London

Colon and Rectal Surgery. Marvin L Corman Lippincott- Raven 1998

Principles and Practice of Surgery for the Colon, Rectum and Anus . Philip H Gordon and Santhat Nivatongs Quality Medical Publishing 1999

The Principles and Practice of Rectal Surgery. William B Gabriel H.K.Lewis and Co London 1963

Surgery of the Anus, Rectum and Colon. Michael R B Keighley and Norman S Williams. W B Saunders 1993

CPSIA information can be obtained at www.ICGtesting.com
Printed in the USA
BVOW07s1210230915

419322BV00007B/123/P